63 Structure and Bonding

Bond and Structure Models

With Contributions by
J. C. A. Boeyens S. N. Bose J. C. Hawes
D. M. P. Mingos K. Nag S.-C. Tam
R. J. P. Williams

With 66 Figures and 39 Tables

Springer-Verlag
Berlin Heidelberg GmbH

ISBN 978-3-662-15255-3 ISBN 978-3-540-39655-0 (eBook)
DOI 10.1007/978-3-540-39655-0

Library of Congress Catalog Card Number 67-11280

Originally published by Springer-Verlag Berlin Heidelberg New York Tokyo in 1985
Softcover reprint of the hardcover 1st edition 1985

2152/3140-543210

Table of Contents

Complementary Spherical Electron Density Model

D. Michael P. Mingos and Jeremy C. Hawes

Inorganic Chemistry Laboratory, University of Oxford, South Parks Road, Oxford OX1 3QR, United Kingdom

The bonding in inorganic molecules of the main group and transition metals is discussed in terms of a model which accounts simultaneously for their stereochemistries and their adoption of the inert gas counting rules. A molecular compound can be viewed initially as a central atom surrounded by a spherical shell of electron density, which is representative of the ligand co-ordination sphere. Since the wave functions for this spherical shell are derived from the particle on a sphere problem it is an easy matter to define the conditions for the inert gas rule in this hypothetical situation, because the wave functions for the sphere and the central atom are both expressed in terms of spherical harmonics with identical quantum numbers. The linear combinations of ligand orbitals in a real complex can be expressed as spherical harmonic expansions and their nodal characteristics defined by the same quantum numbers. Only co-ordination polyhedra where the atoms provide effective coverage or packing on the sphere generate linear combinations in the sequential fashion S, P, D, etc.. These orbitals interact in a complementary fashion with the valence orbitals of the central atom to give a complete set of molecular orbitals, which emulate those of an inert gas in number and nodal characteristics. This Complementary Spherical Electron Density Model thereby provides an effective way of accounting for the stereochemistries of main group and transition metal compounds.

A. Introduction

Understanding the factors which influence the shapes of inorganic molecules represents an on-going challenge for quantum mechanics. The difficulty of providing exact quantum mechanical solutions for chemical systems of interest has meant that this stereochemical problem has had to be approached using approximate methods, none of which has proved to be totally satisfactory. The common occurrence of stable electronic configurations related to those of the inert gases was utilised by Lewis initially and subsequently extended by Sidgwick to formulate the inert gas rule[1, 2]. Although this rule generally defined the stoichiometries of many chemical compounds, it did not provide three dimensional stereochemical descriptions of their structures.

It was Pauling[3] who was able to develop an approximate quantum mechanical model based on the Valence Bond Theory which successfully rationalised the shapes of simple organic, inorganic and co-ordination compounds in terms of hybrid orbitals on the central atom. This localised view of bonding represented a quantum mechanical description of the electron pair bond concept developed by Lewis. Furthermore when combined with the concept of resonance it had a profound and important impact on the development of organic and inorganic chemistry throughout the nineteen forties and fifties[4]. The valence bond method could not be applied in an economical fashion to electron deficient, odd-electron and organo-transition metal π-complexes[5].

The Sidgwick-Powell approach which was extended and popularised by Nyholm and Gillespie in the nineteen fifties[6, 7], focused attention on the total number of electron pairs surrounding the central atom in a molecule. The idea that electron pairs would tend

Table 1a. Summary in a matrix form of the stereochemistries of main group molecular compounds

n_e \ N	7	6	5	4	3	2
8N {2N}	 IF_7	 IOF_5	 SOF_4	 CH_4, SNF_3	 BF_3	 BeF_2
8N + 2 {2N + 2}			 BrF_5	 SF_4	 NH_3, $[IO_3]^-$	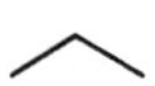 SO_2
8N + 4 {2N + 4}				 XeF_4	 ClF_3	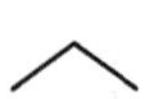 SeH_2, XeO_2
8N + 6						 XeF_2

to occupy regions of space as widely separated as possible had a classical pictorial simplicity and led to a number of preferred co-ordination polyhedra. This approach emphasised the fact that the shapes of main group inorganic molecules could be understood in terms of a small number of easily defined rules despite the apparent complexity of the quantum mechanical problem. For example, since there is a relationship between the number of electron pairs and deltahedral co-ordination polyhedra the shapes of simple inorganic molecules can be represented in a matrix form such as that shown in Table 1a[8]. Particularly noteworthy is the manner in which lone pairs successively replace vertices of the parent polyhedra when they exceed the number of ligands. Some theoretical justification for the approach has been obtained from the Pauli Exclusion principle[9], but the proposed relative magnitudes for the electron repulsion effects have not found support in modern quantum mechanical calculations[10]. Furthermore, the method is not readily applicable to transition metal co-ordination compounds (Table 1b), sandwich compounds and polyhedral inorganic molecules.

As a result of the availability of high speed computers the molecular orbital method has been used with increased frequency to solve stereochemical problems. The most popular approach for defining the preferred ground and excited state geometries depends on taking sections through the multi-dimensional potential energy surface and examining how the orbital energies and total energies vary as a function of distortion co-ordinates[11]. This Walsh diagram methodology[12] has proved to be widely applicable and reliable and in recent years has been most effectively popularised by Hoffmann and his coworkers[13]. The reasons why the method works so well even when used in conjunction with crude semi-empirical calculations has puzzled theoretical chemists and suggests the presence of an underlying fundamental principle[14]. In spite of its notable successes this approach has

Table 1b. Summary of the stereochemistries of transition metal molecular compounds

n_e \ N	7	6	5	4	3	2
18	$[Mo(CN)_7]^{5-}$	$Cr(CO)_6$	$Fe(CO)_5$ $[Ni(CN)_5]^{3-}$	$Ni(CO)_4$		
16			$Cr(CO)_5$	$[Ni(CN)_4]^{2-}$	$Pt(PPh_3)_3$	
14				$Cr(CO)_4$	$[Rh(PPh_3)_3]^+$	$[Ag(NH_3)_2]^+$
12					$Cr(CO)_3$	

pedagogical limitations since it requires the consideration of many distortion co-ordinates before sensible choices concerning the relative stabilities of alternative geometries can be made.

Above we have noted the historical development of the more important approaches to the stereochemical problem. References 15 and 16 summarise some alternative approaches which have been described. Although familiarity with any one of the major approaches described above can make them into powerful tools for the chemist, pedagogically the situation is far from satisfactory. To the simple and fundamental chemical question "What determines the shapes of molecules?" there is no intellectually satisfactory answer. The valence bond method associates preferred geometries to superior overlap effects, the valence shell electron pair repulsion theory depends on maximising the distances between electron pairs and the Walsh methodology depends on very specific arguments associated with energy changes of specific molecular orbitals along the distortion co-ordinate.

In this review we develop a new stereochemical model, – *The Complementary Spherical Electron Density Model.* This model views molecules as interpenetrating spheres of electron density, and attempts to unify the inert gas rule and molecular orbital formalisms. It depends critically on the ideal that the wave functions of ligands can be expressed in terms of spherical harmonics. This idea can be traced back to a paper by Verkade et al.[17)], but the recent analysis by Stone of cluster compounds using this methodology was particularly formative[18)]. The importance of this concept has also been noted by Quinn[19)], who has described a useful method of illustrating spherical, vector spherical and tensor spherical harmonic functions as projections.

B. Spherical Harmonic Representations of Atomic Orbitals

I. General Mathematical Considerations

The use of hydrogen-like wavefunctions to describe the electronic properties of atoms is widespread and generally accepted. The wavefunctions are represented as the product of a radial $R_{n,l}(r)$ and an angular part $Y_{l,m}$ (θ, ϕ)

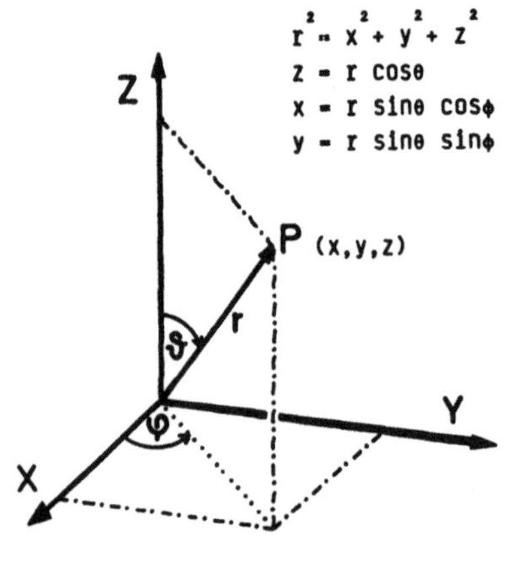

Fig. 1. Definition of spherical polar co-ordinates

$$\Psi(r, \theta, \phi) = R_{n,l}(r) \cdot Y_{l,m}(\theta, \phi) \tag{I}$$

Where r, θ and φ refer to the spherical polar co-ordinate system illustrated in Fig. 1. The angular part is defined by the spherical harmonic functions given in Table 2. For those functions with imaginary solutions the real solutions are obtained by taking linear combinations of complementary components, i.e.

$$Y^c_{l,m} = 1/\sqrt{2}\,[(-1)^m\, Y_{l,m} + Y_{l,-m}]$$

$$Y^s_{l,m} = 1/i\sqrt{2}\,[(-1)^m\, Y_{l,m} - Y_{l,-m}] \qquad \text{For } m > 0 \tag{II}$$

Table 2. Polar forms of the spherical harmonic functions normalised to 4 π

Y_{lm}	Polar Form
Y_{00}	1
Y_{10}	$\sqrt{3}\ \cos\theta$
Y^c_{11}	$\sqrt{3}\ \sin\theta\ \cos\theta$
Y^s_{11}	$\sqrt{3}\ \sin\theta\ \sin\theta$
Y_{20}	$\sqrt{(5/4)}\ (3\cos^2\theta - 1)$
Y^c_{21}	$\sqrt{15}\ \cos\theta\ \sin\theta\ \cos\phi$
Y^s_{21}	$\sqrt{15}\ \cos\theta\ \sin\theta\ \sin\phi$
Y^c_{22}	$\sqrt{(15/4)}\ \sin^2\theta\ \cos 2\phi$
Y^s_{22}	$\sqrt{(15/4)}\ \sin^2\theta\ \sin 2\phi$
Y_{30}	$\sqrt{(7/4)}\ (5\cos^3\theta - 3\cos\theta)$
Y^c_{31}	$\sqrt{(21/8)}\ \sin\theta\ (5\cos^2\theta - 1)\ \cos\phi$
Y^s_{31}	$\sqrt{(21/8)}\ \sin\theta\ (5\cos^2\theta - 1)\ \sin\phi$
Y^c_{32}	$\sqrt{(105/4)}\ \cos\theta\ \sin^2\theta\ \cos 2\phi$
Y^s_{32}	$\sqrt{(105/4)}\ \cos\theta\ \sin^2\theta\ \sin 2\phi$
Y^c_{33}	$\sqrt{(35/8)}\ \sin^3\theta\ \cos 3\phi$
Y^s_{33}	$\sqrt{(35/8)}\ \sin^3\theta\ \sin 3\phi$

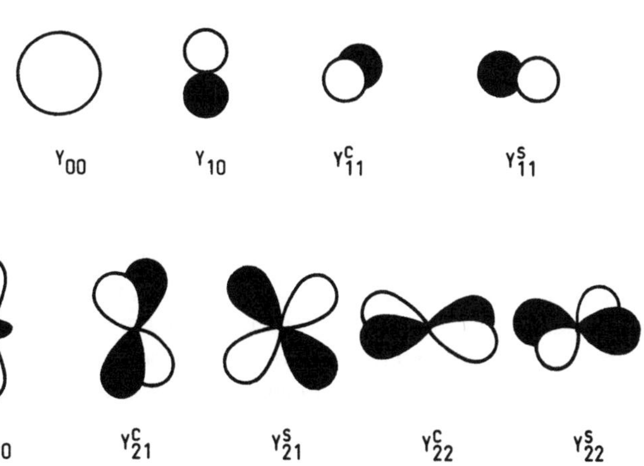

Fig. 2. Illustration of polar plots of the angular parts of the atomic wavefunctions in their real forms

Figure 2 illustrates the angular parts of the wavefunctions in their real forms for s (l = 0), p_z (l = 1, m = 0), p_x (l = 1, m = 1, c) and p_y (l = 1, m = 1, s), d_{z^2} (l = 2, m = 0), d_{xz} and d_{yz} (l = 2, m = 1) and d_{xy} and $d_{x^2-y^2}$ (l = 2, m = 2) atomic orbitals.

The particle on a sphere problem also leads to wave functions which are described in terms of spherical harmonics, but the radial part of the wave function is redundant because the particle is constrained to lie on the surface of the sphere, that is, at a constant radius r. The spherical harmonic solutions are governed by the same quantum numbers l and m and the resultant wave functions can be designated S_m (l = 0), P_m (l = 1) and D_m (l = 2) in an analogous fashion to that adopted for the hydrogen atom.

A molecule can be viewed to a first approximation as a central atom surrounded by a sphere of electron density which has been localised into distinct regions on the sphere. Indeed such a model is the starting point of crystal field theory, where the potential field of an octahedral arrangement of point charges is derived by localising the electron density in just such a manner[20)]. In the present analysis, the molecules are covalent and therefore a molecular orbital analysis is more appropriate. In a molecule ML_N the ligand atomic orbitals are expressed as symmetry adapted linear combinations which combine with the atomic orbitals of the central atom. It is not generally realised, but is very important from the point of view of the present analysis, that it is not necessary to use symmetry arguments to derive ligand linear combinations. As long as the N atoms are distributed in a spherical fashion about the central atom then the linear combinations can be derived to a first approximation from a spherical harmonic expansion. For example, in MH_N if the hydrogen atoms are distributed spherically about M then the symmetry adapted linear combinations $\psi_{l,m}$ of hydrogen 1s functions, σ_i, can be expressed in terms of the following spherical harmonic expansion:

$$\psi_{l,m} = \sum_i c_i \sigma_i \qquad \text{(III)}$$

$$= N' \sum_i Y_{l,m}(\theta_i, \phi_i) \cdot \sigma_i \qquad \text{(IV)}$$

$$= L^\sigma_m \qquad m = 0, 1c, 1s.. \qquad L = S, P, D...$$

where θ_i and ϕ_i represent the locations of the hydrogen nuclei in spherical polar co-ordinates (see Fig. 1) and N′ is a normalising constant. In this fashion the linear combinations are assigned quantum numbers l and m which are related to those which have been defined previously for S, P and D functions derived for the particle on the sphere problem. Furthermore, their nodal characteristics mimic those of the atomic wave functions of the central atom, M. The designation of l and m quantum numbers to symmetry adapted linear combinations of ligand orbitals can be traced back to Verkade et al.[17)]. It has been formulated in a general mathematical fashion and extended to cluster compounds by Stone[18)]. The spherical harmonic expansion described above will provide its most accurate description of the symmetry adapted linear combinations when the ligand polyhedron is a Platonic solid, i.e. tetrahedral, cubic, octahedral, etc. because in those circumstances the polyhedral vertices are symmetry equivalent. It will improve as an approximation as N increases. For smaller polyhedra it is more of an approximation. When there is more than one linear combination with the same symmetry then there can in addition be normalisation problems. The following section describes the utilisation of these spherical harmonic expansions for a range of ligand co-ordination geometries in order to evaluate the types of S^σ, P^σ, D^σ and F^σ functions generated.

II. Planar MH_N Stereochemistries

Figure 3 illustrates schematically the linear combinations of atomic orbitals for some planar H_N aggregates, together with their Stone designations. Clearly these are non-spherical, but they demonstrate the way in which the S^σ, P^σ and D^σ functions develop as N increases. The linear H_2 moiety has a symmetric, S^σ, and antisymmetric, P^σ_{1c}, pair of linear combinations. The H_3 moiety has S^σ and a pair of singly noded P^σ_{1s} and P^σ_{1c} functions. For H_4 the additional function generated is a D function, i.e. it is characterised by S^σ, P^σ_{1s}, P^σ_{1c} and D^σ_{2s}. Therefore, for planar aggregates of atoms not all of the spherical harmonic functions are utilised. The following functions are systematically excluded:
(1) Those which possess a nodal plane coincident with the xy plane of the polygon, viz. P^σ_0, D^σ_{1s}, F^σ_{2c} ... etc..
(2) Those having a form in the xy plane, which when renormalised is identical to those of spherical harmonic functions with lower l, m quantum numbers, e.g. D^σ_0 and S^σ, F^σ_{1c} and P^σ_{1c} are identical in projection.

The remaining allowed set of two dimensional combinations are S^σ, P^σ_{1c}, P^σ_{1s}, D^σ_{2c}, D^σ_{2s}, F^σ_{3c}, F^σ_{3s} ..... $L^\sigma_{l,s}$ are illustrated in Fig. 4. They correspond to the solution of the Schrödinger equation for the particle on the ring problem, i.e. they are the two dimen-

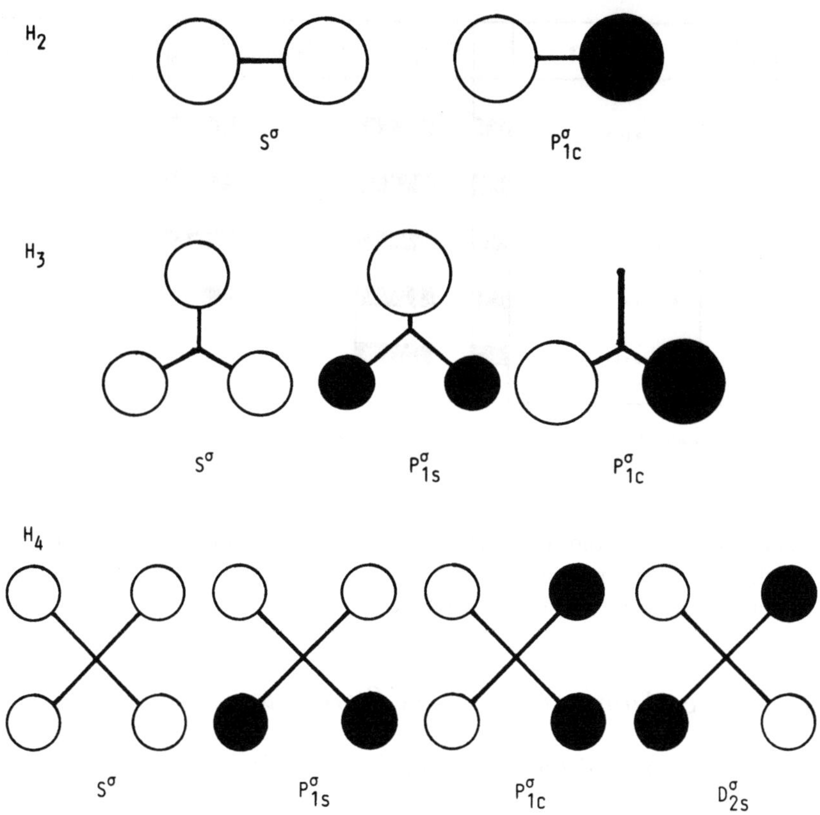

Fig. 3. Schematic representations of the linear combinations, L^σ, for planar aggregates of hydrogen atom 1s functions

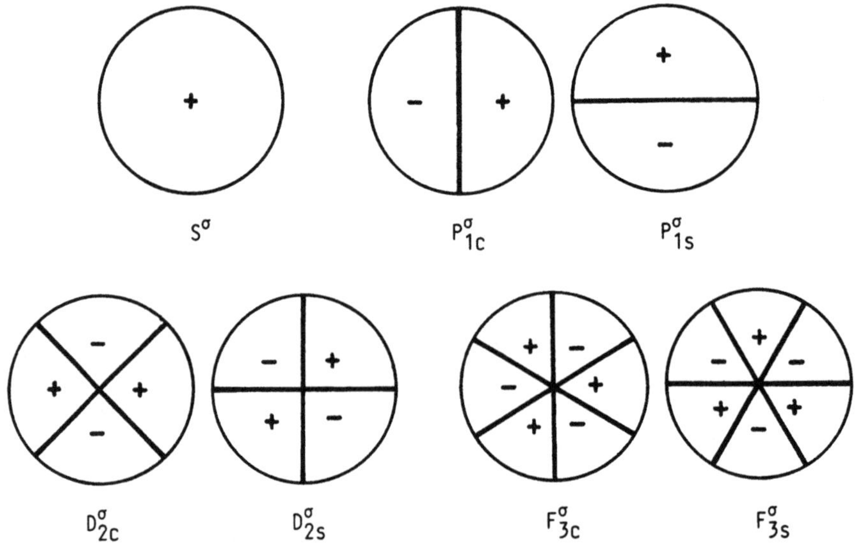

Fig. 4. Schematic representations through the xy plane of the L^{σ} functions representing a two dimensional system

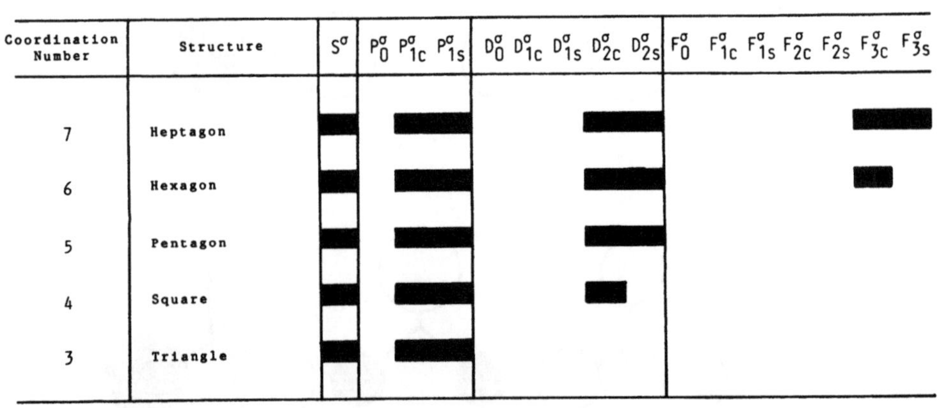

Fig. 5. Summary of linear combinations for planar σ-donor aggregates and their L^{σ}_{m} designators. Particularly noteworthy is the use of F^{σ} functions when $N > 5$

sional analogues of the spherical harmonics. Figure 5 illustrates the stepwise manner in which these functions are utilised for planar H_N aggregates. Particularly noteworthy is the utilisation of F functions when N is $\geqslant 6$.

III. Three Dimensional MH_N Stereochemistries

a. Three Connected Polyhedra and Bipyramids

The simplest three dimensional polyhedron is the tetrahedron and the four linear combinations of atomic orbitals (S^{σ}, P^{σ}_{0}, P^{σ}_{1s}, P^{σ}_{1c}) generated from Eq. IV are illustrated in

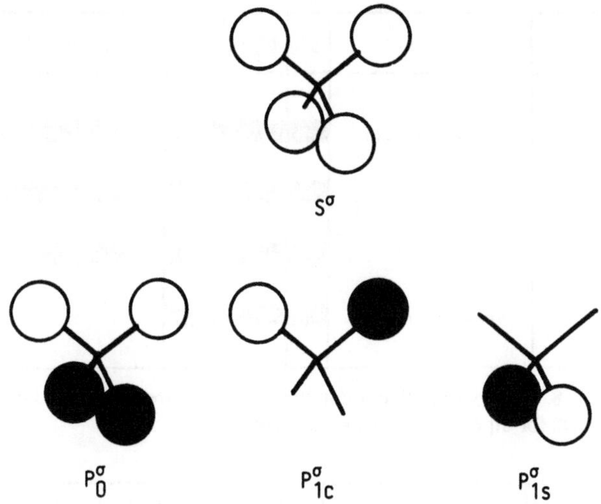

Fig. 6. Linear combinations of 1 s orbitals for tetrahedral H_4

Fig. 6. When N exceeds four then D^σ and even F^σ linear combinations are generated. The particular functions generated depends on the positions of the hydrogen atoms relative to the nodal cones and planes of the D and F functions. In this section the spherical harmonic generated linear combinations for bipyramids and three-connected co-ordination polyhedra are explored. The discussion of planar stereochemistries given above has demonstrated that a P_0^σ combination requires the presence of more than one plane of atoms perpendicular to the principle axis. Similarly a D_0^σ function requires at least three planes, an F_0^σ function four planes or more (see Fig. 7). With this property in mind it is possible to understand the occurrence and absence of the L_0^σ linear combinations of ligand orbitals illustrated in Fig. 8 for three connected polyhedra and in Fig. 9 for bipyramids.

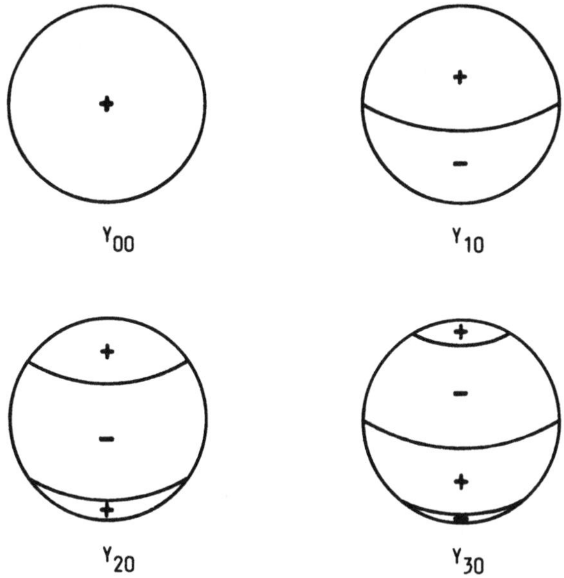

Fig. 7. Nodal characteristics of the cylindrically symmetric $Y_{l,0}$ spherical harmonics

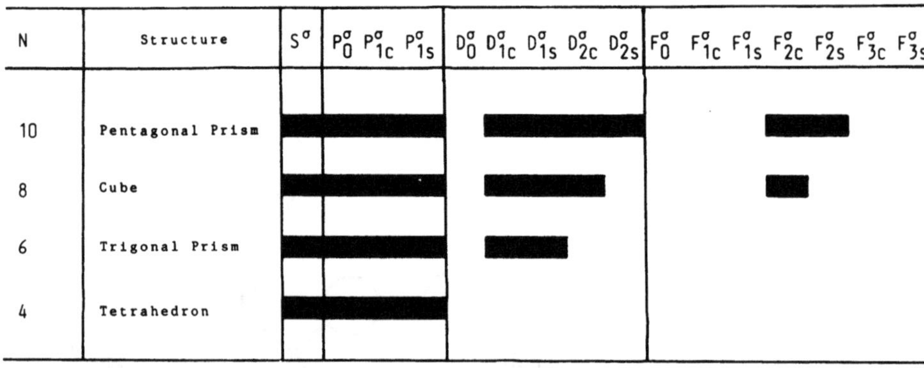

Fig. 8. Summary of linear combinations for three connected co-ordination polyhedra. The absence of D^σ_0 for all examples is particularly noteworthy

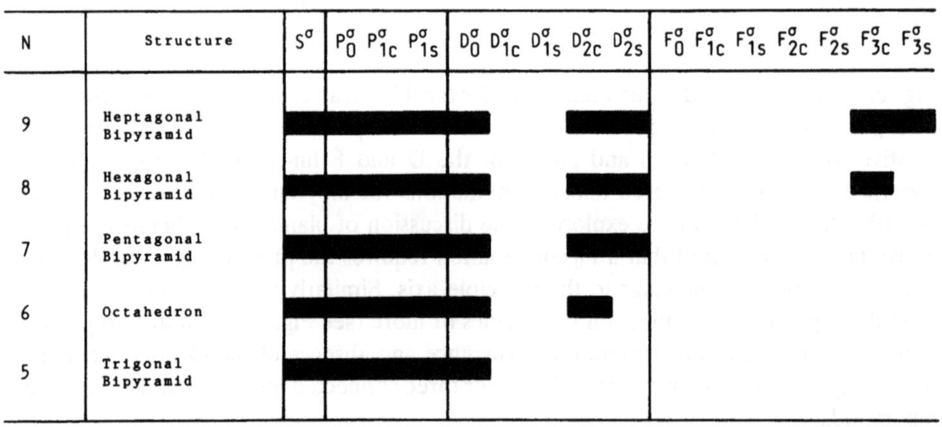

Fig. 9. Summary of linear combinations for bipyramids. D^σ_{1c} and D^σ_{1s} are consistently absent

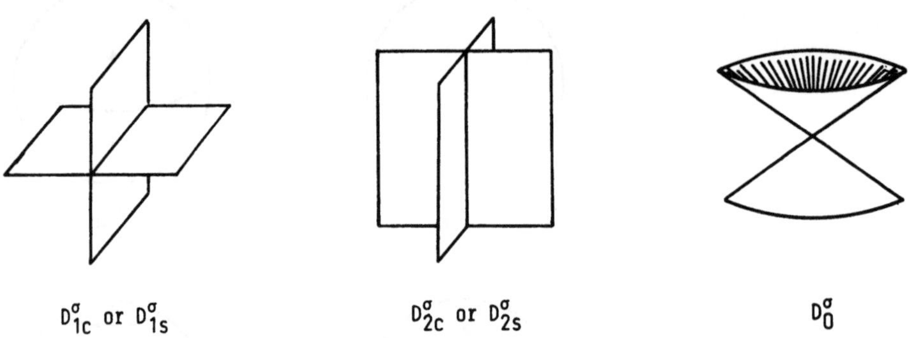

Fig. 10. Nodal characteristics of D^σ functions

F^σ linear combinations are generated for MH_N (N = 5–9) when all the hydrogen atoms lie on the nodal cones of (10 – N) D or P spherical harmonic functions. For example when N = 9, (10 – N) = 1, so at least one D^σ function must be excluded if F^σ are generated. See Fig. 10 for an illustration of the nodal characteristic of the D^σ functions. Location of the atoms along the nodal cones of D_0^σ nullifies this function (all the coefficients, c_i, are equal to 0) and requires the utilisation of an F^σ function. Similarly location of atoms on the nodal planes of either the $D_{1c,s}^\sigma$ or $D_{2c,s}^\sigma$ functions nullifies them and forces the adoption of F^σ functions with an additional nodal plane. See Fig. 11 for a similar analysis of $N = 8$ and $N = 7$ polyhedra, where two and three L^σ functions are nullified because of the location of atoms on nodal lines and planes. For example for bipyramids the D_{1c}^σ and D_{1s}^σ functions are always absent, because bipyramids place the equatorial atoms in the horizontal nodal plane of D_{1s}^σ and D_{1c}^σ while the axial atoms always reside in the vertical nodal planes of these functions.

In summary the three-connected polyhedra and bipyramids with $N > 7$ require the utilisation of F^σ functions in order to account fully for the N linear combinations of atomic orbitals of the peripheral atoms.

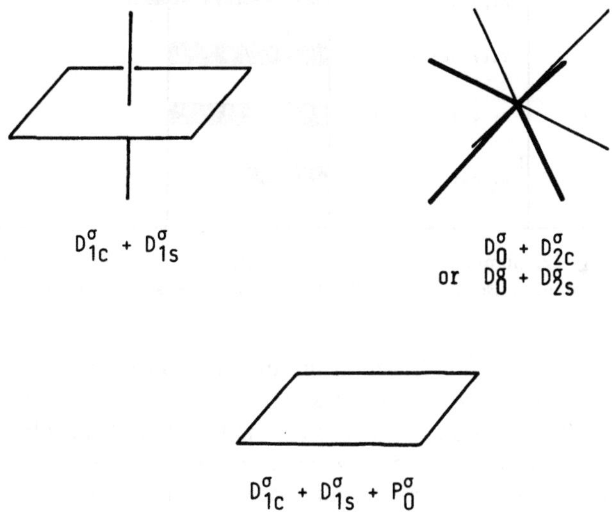

Fig. 11. The nodal characteristics which are common to sets of D^σ and P^σ functions

b. Deltahedra

Polyhedra with triangular faces exclusively are described as deltahedra and have several interesting geometric properties[21]. For example, their vertices are connected by the maximum number of edges. This is a property of importance for the borane polyhedra $B_nH_n^{2-}$, which are electron deficient and require the maximum degree of delocalisation in order to stabilise the boron skeleton[18]. Figure 12 gives a detailed analysis of the development of the linear combinations of atomic orbitals for deltahedra and classifies them according to their l, m quantum numbers. Particularly striking is the sequental fashion in which the linear combinations are built up; unlike the previous classes of polyhedra F^σ functions are not used prior to the completion of the D shell. This property can be related to the ability of deltahedra to give the best coverage on a sphere, i.e. the most even distribution of points. Consequently, the polyhedra reproduce most closely the surface of the parent spherical shell which surrounds the central atom in ML_N.

Mathematically this property has been explored previously in the context of the following covering problem by Fejes-Toth and others[22].

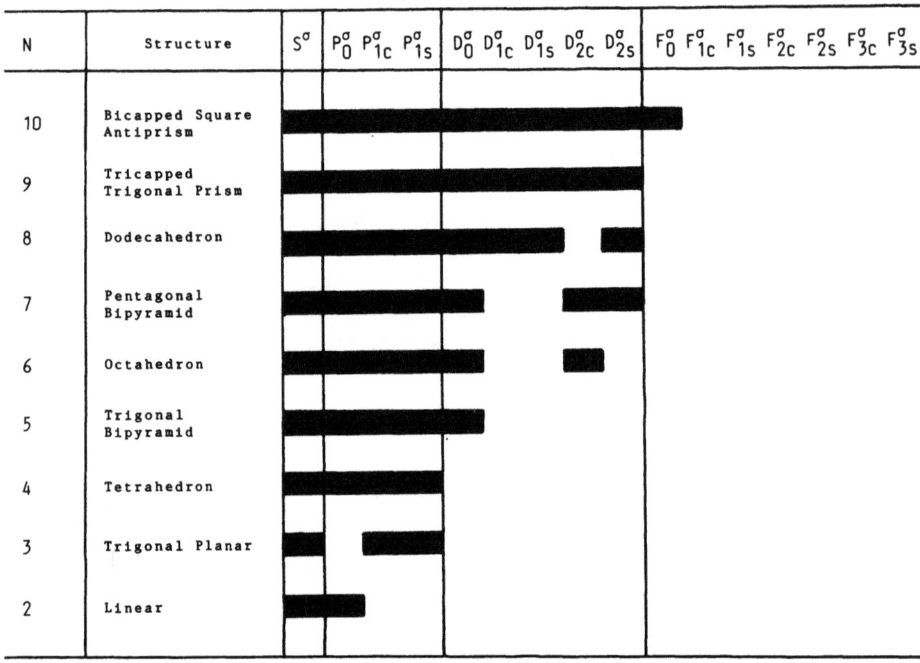

Fig. 12. Summary of linear combinations for deltahedral aggregates of ligand σ-orbitals. The sequential use of S^σ, P^σ, D^σ, and F^σ functions is noteworthy

If N oil supply depots are available on the surface of the sphere what is their best arrangement to give the most efficient utilisation of oil resources?

The solution to this problem is based on deltahedral arrangements of points, since circles from these points which overlap and thereby give complete coverage are of smaller radius than those for alternative polyhedral arrangements. In a chemical context a

Table 3. Symmetry designations for the ligand linear combinations and central atom orbitals in deltahedral co-ordination compounds

	N	2	3	4	5	6	7	8	9	10
	Point Group	$D_{\infty h}$	D_{3h}	T_d	D_{3h}	O_h	D_{5h}	D_{2d}	D_{3h}	D_{4d}
<u>L_N</u>	<u>M</u>									
S^σ	s	a_{1g}	a_1'	a_1	a_1'	a_{1g}	a_1'	a_1	a_1'	a_1
P_0^σ	p_z	a_{1u}	a_2''		a_2''		a_2''	b_2	a_2''	b_2
P_{1c}^σ	p_x	e_{1u}	e'	t_2	e'	t_{1u}	e_1'	e	e'	e_1
P_{1s}^σ	p_y									
D_{1c}^σ	d_{xz}	e_{1g}	e''		e''		e_1''	e	e''	e_3
D_{1s}^σ	d_{yz}			t_2		t_{2g}				
D_{2s}^σ	d_{xy}	e_{2g}	e'		e'		e_2'	b_2	e'	e_2
D_{2c}^σ	$d_{x^2-y^2}$			e		e_g		b_1		
D_0^σ	d_{z^2}	a_{1g}	a_1'		a_1'		a_1'	a_1	a_1'	a_1

deltahedral arrangement of ligands on a sphere provides a collection of overlapping regions of electron density which most effectively approximate to a spherical shell.

The symmetry properties of the linear combinations of atomic orbitals for deltahedra are summarised in Table 3 and the following characteristics are particularly noteworthy.

The linear combinations have identical symmetry characteristics to those of atomic orbitals located on the central atom and with matching l and m quantum numbers. This is a direct consequence of definining both sets in terms of spherical harmonics.

Only for the Platonic solids do S^σ, and D_0^σ functions have different symmetry properties. This is related to the occurence of cubic fields for these high symmetry polyhedra.

For the polyhedra with T_d, D_{3h} and D_{2d} symmetries some of the P^σ and D^σ functions have identical symmetry characteristics, e.g. $P^\sigma(t_2)$ and $D^\sigma(t_2)$ in T_d. Therefore, although the p and d atomic orbitals for an isolated atom are orthogonal, the corresponding linear combinations $P^\sigma(t_2)$ and $D^\sigma(t_2)$ can mix. This arises because the peripheral atoms do not define a perfect sphere and the designation of l, m quantum numbers is only an approxi-

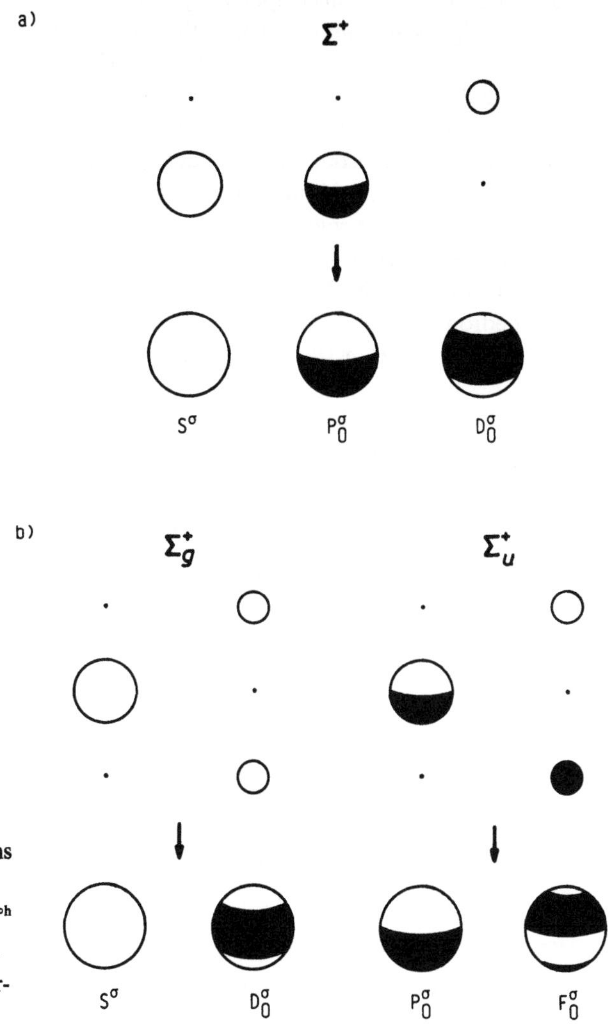

Fig. 13 a, b. Schematic illustrations of (**a**) mono- and (**b**) bicapping of polyhedral aggregates. $C_{\infty v}$ and $D_{\infty h}$ symmetry is assumed to define the symmetry labels. Either one (**a**) or two (**b**) new L_0^σ functions are generated respectively mate one. Nevertheless, pseudo-symmetry considerations will serve to effectively limit the degree of mixing between P and D functions.

c. *Capped Polyhedra*

Co-ordination polyhedra can at times be described as capped polyhedra, e.g. 7 atoms can form an octahedron with the seventh atom located on a face. If the capping atom is introduced along the z axis the polyhedral L^σ linear combinations can be defined in terms of $C_{\infty v}$ symmetry labels. The additional capping orbital has Σ^+ symmetry and therefore gives rise to the formation of an additional Σ^+ linear combination (Fig. 13 a)[2]. Some examples of this process are illustrated in Fig. 14. For the octahedron and square-antiprism this results in the generation of an additional D_0^σ function. Trans-bicapping results in the generation of D_0^σ and F_0^σ functions corresponding to in-phase, Σ_g^+, and out-of phase Σ_u^+, linear combinations of the capping orbitals interacting out of phase with the parent S^σ and P_0^σ polyhedral orbitals respectively (see Fig. 13 b, where the effective symmetry is $D_{\infty h}$). From Fig. 14 it is apparent that the capped octahedron and capped square-antiprism provide a set of S^σ, P^σ and D^σ functions which correspond to those derived previously for deltahedra, i.e. they do not utilise out-of sequence F functions. Consequently, they emulate the parent spherical shell effectively.

This brings us to a second mathematical property associated with spheres, described by mathematicians as a packing problem. It is usually stated in the following fashion:

If N inimical dictators control the planet how could they be located on the surface of the sphere so as to maximise the distances between them? The following solutions to this problem have been derived by mathematicians[22]:

4 tetrahedron
5 trigonal bipyramid or square pyramid

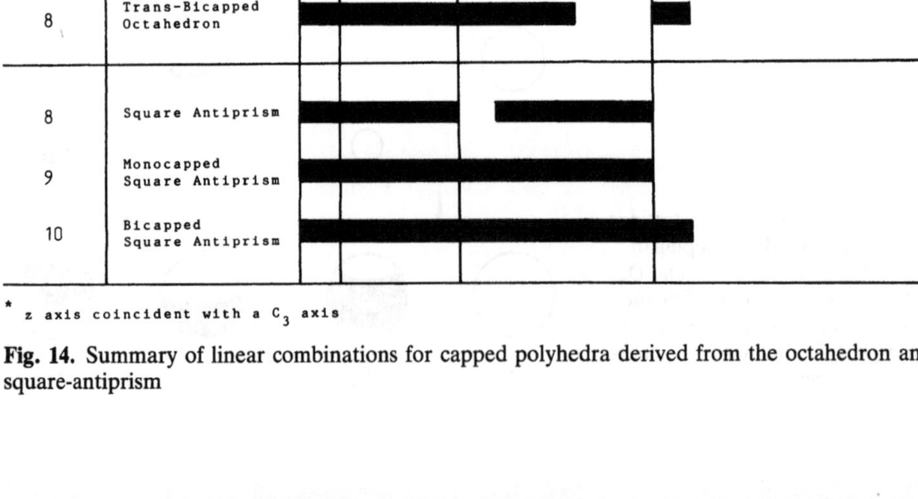

* z axis coincident with a C_3 axis

Fig. 14. Summary of linear combinations for capped polyhedra derived from the octahedron and square-antiprism

6 octahedron
7 capped octahedron
8 square-antiprism
9 capped square-antiprism
10 bicapped square-antiprism
12 icosahedron.

In a chemical context many calculations of this sort have been performed because this packing problem is closely related to that of finding the disposition of electron pairs which maximises their mutual separations[24].

What is important in the present context is that solution either to the covering or packing problems associated with the sphere gives rise to ligand polyhedra which can be represented by spherical harmonic expansions which most closely follow those of a spherical shell. Particularly noteworthy is the fact that although the cube has higher symmetry than either the square antiprism or the dodecahedron it does not generate a set of four D^{σ}_{-} functions alone because the points of a cube do not provide an effective coverage of the sphere.

C. Mathematical Formulation of the Inert Gas Rule

In the previous section the idea of expressing linear combinations of ligand orbitals as a spherical harmonic expansion was introduced. The consequences of this simple idea for describing the bonding in ML_N complexes and the electronic factors responsible for the adoption of the inert gas rule by low valent transition metal complexes and main group molecules will be developed in this section. It is striking that although the inert gas rule has been widely used by chemists for almost fifty years there have been only a handful of papers dealing with its theoretical basis. In particular only Craig attempted a general formulation[25], while the majority of papers and textbooks have limited its theoretical basis to examples based on d^6 octahedral complexes[25–30]. Prior to a detailed discussion it is instructive to consider the hypothetical situation of a central atom with its associated valence orbitals surrounded by a spherical shell of electron density. This serves as a model for the ligand co-ordination sphere (see Fig. 15). The latter has wave functions which are expressed as spherical harmonics with characteristic l, m quantum numbers and designated S^{σ}, $P^{\sigma}_{mc,s}$, and $D^{\sigma}_{mc,s}$. The functions are not degenerate and their energies increase in the order $S < P < D$, i.e. they follow the nodal characteristics of the functions. If interactions between the central atom wave functions, $\phi_{l,m}$, and the spherical shell wave functions, $\psi_{l,m}$, occur they can be represented in a molecular orbital framework by taking linear combinations in the following fashion:

$$\psi^{+}_{l,m} = N(\psi_{l,m} + \lambda\phi_{l,m}) \tag{V}$$

representing the bonding combination. In this spherically symmetric situation orthogonality relationships limit the expansion to include only those functions with the identical l, m quantum numbers, i.e.

$$\langle\psi_{l,m}|\hat{\mathcal{H}}|\phi_{l',m'}\rangle = 0 \quad \text{unless} \quad l = l' \quad \text{and} \quad m = m' \tag{VI}$$

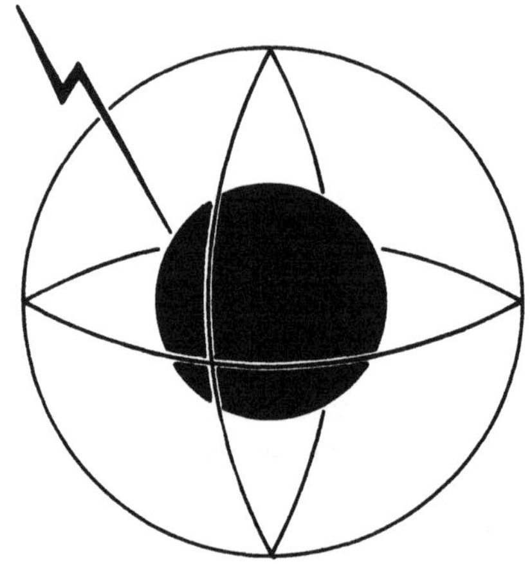

Fig. 15. Idealised representation of a molecule as a central atom surrounded by a sphere of electron density

Therefore bonding molecular orbitals arise from the pairwise interaction of matching functions on the central atom and the ligand spherical shell. The number and quantum numbers of the bonding molecular orbitals are determined by the valence orbitals of the central atom. For example if the central atom has s, p_z, p_x, and p_y valence orbitals then they overlap in a pairwise fashion with S^σ, P_0^σ, P_{1c}^σ and P_{1s}^σ spherical shell functions to form four bonding molecular orbitals which can accomodate a total of 8 electrons. The resultant electron distribution is spherical since both $p_x^2 + p_y^2 + p_z^2$ and $P_0^2 + P_{1s}^2 + P_{1c}^2$ are spherically symmetric. Furthermore ionization from the resultant bonding molecular orbitals will require considerable energy if the resonance integrals $\langle \psi_{l,m}|\hat{\mathcal{H}}|\phi_{l,m}\rangle$ are large. Additional electrons could be accomodated on the spherical shell in D^σ functions, but these are higher lying than the S^σ and P^σ functions because of the additional number of nodes. Furthermore, unless all the D^σ functions are filled the electron distribution is not spherical.

If the central atom has d, s and p valence orbitals which have large resonance integrals with the spherical shell D^σ, P^σ, and S^σ functions then a total of nine stable molecular orbitals are generated. The F^σ functions on the ligand sphere are non-bonding with respect to the central atom and higher lying than the original S^σ, P^σ and D^σ functions.

For this idealised model adherence to the inert gas rule has two prerequisites: a spherical distribution of electron density arising from matching orbitals on the spherical shell and central atom, and large resonance integrals between the two sets of orbitals which lead to molecular orbitals with high ionisation potentials. The correspondence between these molecular orbitals and the atomic orbitals of the inert gas atom are transparently obvious for this hypothetical model. In real molecules the atoms adopt stereochemistries which emulate this ideal as most effectively as possible.

In a real ML_N molecule the ligand spherical shell is concentrated into N distinct regions of electron density and there are no longer a limitless number of spherical harmonic functions associated with the ligand sphere. Only N linear combinations of ligand orbitals can be generated and their l, m quantum numbers are determined completely by the distribution of atoms on the ligand sphere in the manner described in detail in the previous section. The orthogonality relationship:

$$\langle \psi_{l,m} | \phi_{l',m'} \rangle = 0 \quad \text{unless} \quad l = l', m = m' \tag{VII}$$

is no longer strictly valid because spherical symmetry is no longer present, but will nonetheless serve as a good approximation because pseudo-spherical symmetry is maintained and the nodal characteristics of the linear combinations accurately reflect those of the parent spherical shell. The approximation improves as N gets larger.

For reasons described in the previous section polyhedra which give the best solution to the covering and packing problems on the sphere generate spherical harmonic expansions which most closely emulate those of the spherical shell.

Figure 16 illustrates the generation of the molecular orbitals of CH_4 from the carbon 2s and 2p atomic orbitals and the S^σ and P^σ linear combinations of hydrogen atomic orbitals. The cubic symmetry of the tetrahedron ensures the pairwise interactions of ligand and central atom functions and leads to four stable molecular orbitals, which can be occupied by 8 electrons. No non-bonding or antibonding functions are occupied. The

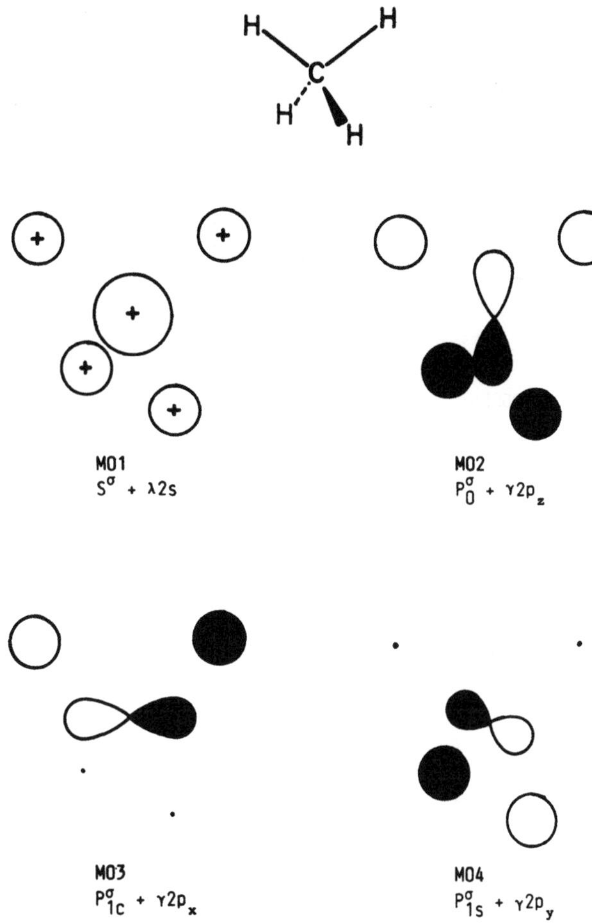

Fig. 16. Bonding molecular orbitals of a_1 (S^σ) and t_2(P^σ) symmetry for CH_4

alternative square-planar geometry is disfavoured because the hydrogen 1s linear combinations – S^σ, P^σ_{1s}, P^σ_{1c} and D^σ_{2s} (see Fig. 3) do not find a complete matching set of central atom orbitals. The empty p_z atomic orbital remains localised on the central atom and the occupied D_{2s} function on the ligand shell. Consequently, only three orbitals are utilised in bonding and the electron distribution is far from spherical[31]. The accuracy of the spherical harmonic description for CH_4 is emphasised by electron density plots of the a_1 and t_2 molecular orbitals derived from ab initio calculations and illustrated in Fig. 17. Particularly noteworthy is their resemblance to the hydrogen 2s and 2p functions also illustrated in the Figure[32].

Figure 18 illustrates the molecular orbitals of $[ReH_9]^{2-}$ derived from Re 5d, 6s and 6p atomic orbitals and the hydrogen S^σ, P^σ, and D^σ linear combinations for a tricapped trigonal prismatic geometry. Once again there is a 1:1 match of ligand and central tom

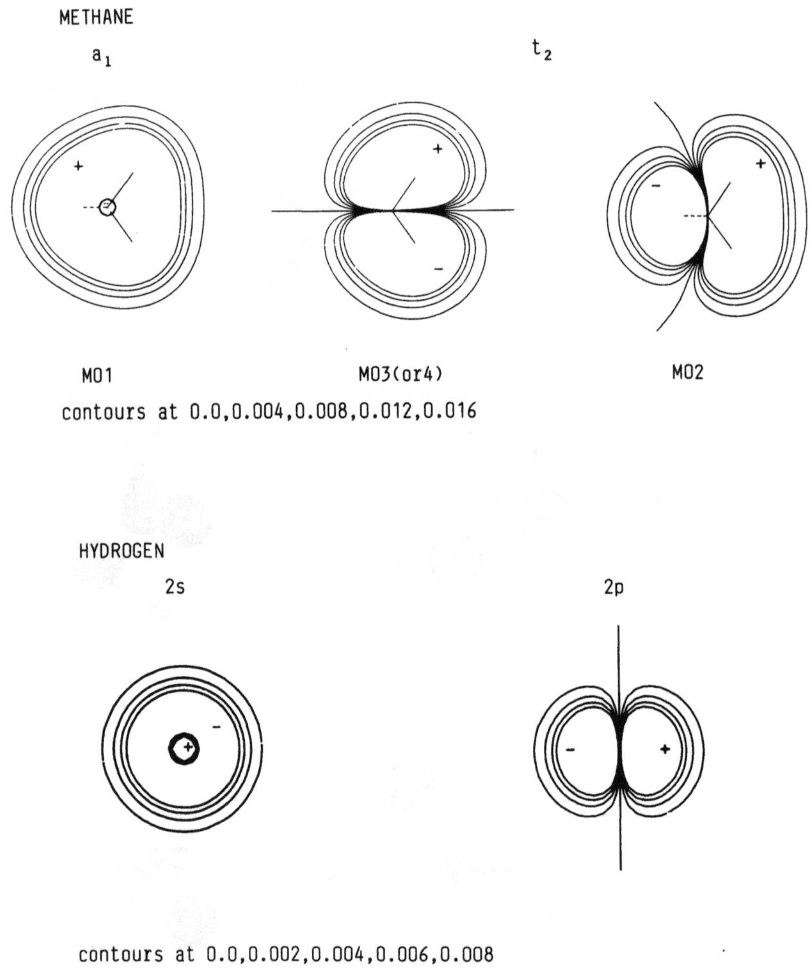

Fig. 17. A comparison of the electron density plots for CH_4 and the 2s and 2p orbitals of the hydrogen atom. (Reproduced from Streitwieser A., Owens, P. H.: Orbital and Electron Density Diagrams, Macmillan, New York, 1973, with permission.)

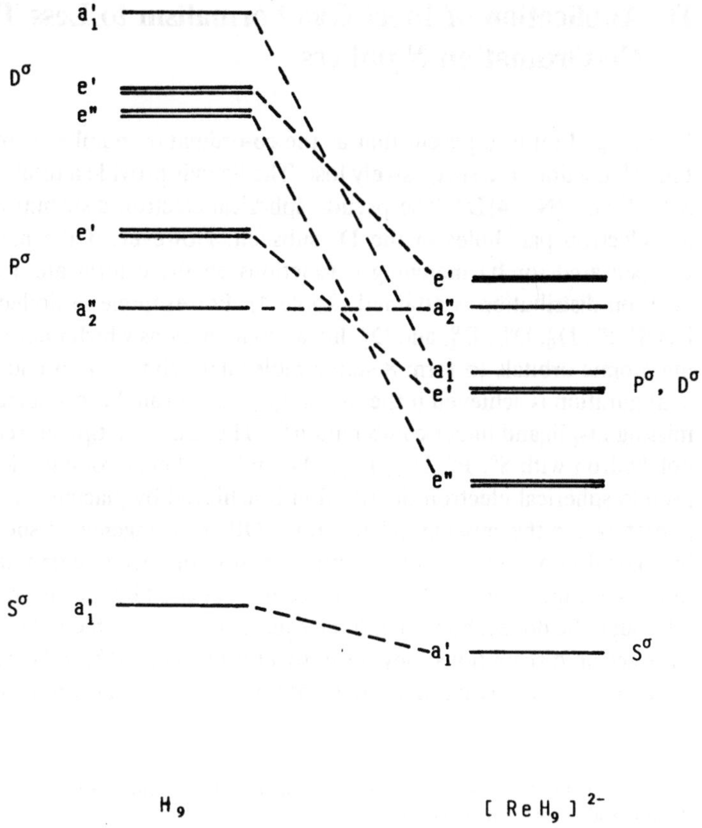

Fig. 18. A schematic representation of the molecular orbital diagram for $[ReH_9]^{2-}$ adapted from Ref. 33. Particularly noteworthy is the stabilisation of the H_9 P^σ and D^σ functions to give a closely spaced band of bonding molecular orbitals

orbitals and the formations of 9 stable molecular orbitals with nodal characteristics which emulate those of the inert gas valence orbitals[33]. A capped square-antiprismatic ligand arrangement also generates a complete set of S^σ, P^σ, and D^σ functions which would provide an alternative stereochemistry which satisfies the requirements of the inert gas formulation (see Fig. 14). However, a heptagonal bipyramid would not provide a suitable stereochemistry because two of the D^σ functions are replaced by F^σ functions, which find no match in the valence orbitals of the central transition metal atom (see Figs. 9 and 11). The occurrence of two alternative polyhedra which lead to approximately spherical electron distributions based on S, P and D functions is consistent with the observed stereochemical non-rigidity of $[ReH_9]^{2-}$ [34]. In summary the inert gas rule has been reformulated in terms of complementary sets of strongly interacting ligand and metal functions which generate as closely as possible a spherical, stable and complete set of S^σ, P^σ, and D^σ wave functions. The great advantage of this reformation is that it carries with it stereochemical implications because only those polyhedra which provide an effective covering or packing of the sphere generate the appropriate linear combinations of orbitals.

D. Application of Inert Gas Formalism to Less Than Nine Co-Ordination Numbers

From Fig. 12 it is apparent that as the co-ordination number is reduced from nine then the D^σ functions are successively lost. The ligands provide a total of N $L^\sigma_{m,l}$ functions, S^σ, $3 \times P^\sigma$ and $(N-4)D^\sigma$. The pseudo-spherical electron distribution is lost because there are electron pair holes in the D^σ subshell. However, if the missing D^σ functions are compensated for by matching d functions on the central atom then a more spherical electron distribution is attained (Table 4). For example, a dodecahedral MH_8 complex has S^σ, P^σ, D^σ_0, D^σ_{1c}, D^σ_{1s} and D^σ_{2s} linear combinations which interact strongly with matching atomic orbitals to form 8 stable molecular orbitals. A pseudo-spherical 18 electron configuration is achieved if the metal $d_{x^2-y^2}$ (d_{2c}) orbital is occupied in order to match the missing D^σ_{2c} ligand linear combination[35]. The square-antiprism represents an alternative polyhedron with S^σ, P^σ, D^σ_{1c}, D^σ_{1s}, D^σ_{2c} and D^σ_{2s} linear combinations (see Fig. 19) and a pseudo-spherical electron distribution is achieved by placing a electron pair in d_{z^2} compensating for the missing D^σ_0 function. Other arrangements such as the cube and the hexagonal bipyramid are disfavoured because the ligand combinations include F^σ functions (see Figs. 8, 9 and 11) which are not matched by the central atom wavefunctions. Although the dodecahedron and the square-antiprism achieve 18 electron pseudo-spherical electron distributions they do so by utilising a metal based orbital which in general is going to be higher lying than the eight bonding molecular orbitals. Clearly those ligands

Table 4. Symmetry designations for the ligand linear combinations and central atom orbitals in deltahedral co-ordination compounds

N / Point Group	8 D_{2d}	8 D_{4d}	8 C_{2v}	7 D_{5h}	7 C_{2v}
Structure	Dodecahedron	Square antiprism	Bicapped trigonal prism	Pentagonal bipyramid	Tetragonally capped trigonal prism
Ligand field splitting & Occupancy for 18e species	b_2 — xy e = xz, yz a_1 — z^2 b_1 –θ– x^2-y^2	e_3 = xz, yz e_2 = xy, x^2-y^2 a_1 –θ– z^2	b_1 — xz b_2 — yz a_1 — x^2-y^2 a_2 — xy a_1 –θ– z^2	a_1' — z^2 e_2' = xy, x^2-y^2 e_1'' =θθ= xz, yz	a_2 — xy a_1 — z^2 b_1 — xz b_2 –θ– yz a_1 –θ– x^2-y^2
Associated ligand sphere D^σ functions	D^σ_0 D^σ_{1c} D^σ_{1s} D^σ_{2s}	D^σ_{1c} D^σ_{1s} D^σ_{2c} D^σ_{2s}	D^σ_0 D^σ_{1c} D^σ_{1s} D^σ_{2s}	D^σ_0 D^σ_{2c} D^σ_{2s}	D^σ_0 D^σ_{1c} D^σ_{2s}

7 C_{3v}	6 O_h	6 D_{3h}	5 D_{3h}	5 C_{4v}	4 T_d
Monocapped octahedron	Octahedron	Triangular prism	Trigonal bipyramid	Square pyramid	Tetrahedron
e = xz, yz a_1 — z^2 e =θθ= xy, x^2-y^2	e_g = z^2, x^2-y^2 t_{2g} ≡θθθ≡ xy, xz, yz	e'' = xz, yz e' =θθ= xy, x^2-y^2 a_1' –θ– z^2	a_1' — z^2 e' =θθ= xy, x^2-y^2 e'' =θθ= xz, yz	b_1 — x^2-y^2 a_1 –θ– z^2 b_2,e ≡θθθ≡ xy, xz, yz	t_2 ≡θθθ≡ xy, xz, yz e =θθ= z^2, x^2-y^2
D^σ_0 D^σ_{1c} D^σ_{1s}	D^σ_0 D^σ_{2c}	D^σ_{1c} D^σ_{1s}	D^σ_0	D^σ_{2c}	None

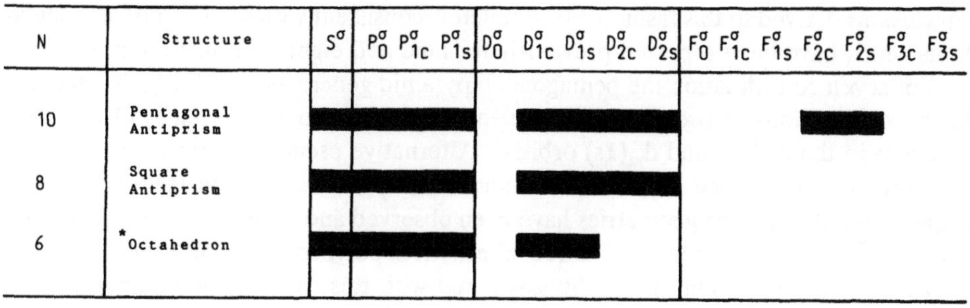

* z axis coincident with a C_3 axis

Fig. 19. Linear combinations for antiprismatic co-ordination polyhedra

which show appreciable covalent interactions give a set of molecular orbitals whose radial characteristics approximate to those of the next inert gas atom and are effective in giving complexes which adhere to the inert gas rule. As illustrated for CH_4 (Fig. 17) hydrogen is an effective ligand in this context. When it bonds to a transition metal it is able to overlap well with the contracted nd orbitals, and its S^σ and P^σ linear combinations are able to emulate the metal 4s and 4p atomic orbitals. The non-bonding d function localised on the metal can also be stabilised by π-acceptor ligands. These π-bonding effects will be discussed in more detail subsequently[36]. Examples of nine and eight co-ordinate complexes which conform to the eighteen electron rule are summarised in Table 5 and 6[37]. Also, given in the Tables are examples of eight co-ordinate complexes with 17 and 16 electrons corresponding to partial and non-occupation of the metal based orbital. The

Table 5. Some examples of nine co-ordinate complexes

Complex	Structure
$[Ho(H_2O)_9](EtSO_4)_3$	TTP
$[Pr(H_2O)_9](BrO_3)_3$	TTP
$[ReH_9]^{2-}$	TTP
$[TcH_9]^{2-}$	TTP
$Tl(NO_3)_3(H_2O)_3$	TTP
$Th(CF_3COCHCOCH_3)_4(H_2O)$	CSAP

Table 6. Some examples of eight co-ordinate complexes

Complex	d^n	Structure
$La(acac)_3 \cdot 3H_2O$	0	BTP
$[ZrF_8]^{4-}$	0	BTP
$K_3Cr(O_2)_4$	1	DD
$NbCl_4(diars)_2$	1	DD
$H_4Mo(PPh_3)_4$	2	DD
$H_4Mo(PMe_2Ph)_4$	2	DD
$H_4W(CN)_8 \cdot 6H_2O$	2	SAP
$K_4W(CN)_8 \cdot 2H_2O$	2	DD

conclusions derived in this fashion are completely consistent with semi-empirical calculations which have been reported previously for nine and eight coordinate complexes[35].

For seven co-ordination the pentagonal bipyramid generates S^σ, P^σ, D_0^σ, D_{2c}^σ and D_{2s}^σ linear combinations. A pseudo-spherical electron distribution is achieved by placing four electrons in the $d_{xz}(1c)$ and $d_{yz}(1s)$ orbitals. Alternative pseudo-spherical situations are achieved for the capped trigonal prism and the capped octahedron (see Table 4)[38]. Experimentally all three geometries have been observed and Table 7 gives some specific examples[39]. The occurrence of three alternative polyhedra is consistent with the stereochemical non-rigidity generally associated with this co-ordination number. Once again in the absence of π-bonding effects the metal based orbital can be depopulated to give 17, 16, 15 and 14 electron complexes with comparable stabilities, but lacking spherical electronic distribution.

Table 7. Some examples of seven co-ordinate complexes

Complexes	d^n	Structure
$[NbF_7]^{2-}$	0	CTP
$[ZrF_7]^{3-}$	0	PB
$MoBr_4(PMe_2Ph)_3$	2	CO
$[V(CN)_7]^{4-}$	2	PB
$H_4Os(PMe_2Ph)_3$	4	PB
$[Mo(CNBu^t)_7]^{2+}$	4	CTP
$W(CO)_4Br_3$	4	CO

An octahedral ligand set gives rise to S^σ (a_{1g}), P^σ (t_{1u}) and D_0^σ, D_{2c}^σ (e_g) linear combinations and a complementary electron distribution is achieved by populating the metal d_{xz}, d_{yz} and d_{xy} (t_{2g}) orbitals, which compensate for the missing D_{1c}^σ, D_{1s}^σ and D_{2s}^σ ligand functions.

The trigonal prism provides a less satisfactory solution to the covering and packing problem, but nonetheless it does generate a set of ligand functions based exclusively on S^σ, P^σ and D^σ functions, i.e. S^σ (a_1'), P^σ (e', a_2''), D_{1c}^σ and D_{1s}^σ (e''). The complementary set of metal d orbitals is d_{z^2} (a_1') and d_{xy}, $d_{x^2-y^2}$ (e'). In contrast to the octahedron the central atom d, s and p orbitals share common irreducible representations of the D_{3h} point group, i.e. s and $d_{z^2} - a_1'$; p_x, p_y and d_{xy}, $d_{x^2-y^2} - e'$[40].

This permits mixing of orbitals with different l, m quantum numbers, and signals a breakdown of the spherical approximation. A particularly important consequence of this is that it allows mixing of the ligand $P_{1s,1c}^\sigma$ and the metal d_{xy} and $d_{x^2-y^2}$ orbitals and destabilises the latter as a consequence[41]. Therefore, in an 18 electron complex the highest occupied molecular orbitals are somewhat metal-ligand antibonding. In terms of the spherical electron density model such mixings can be viewed as perturbations which reduce the spherical symmetry and raise the energy of the system. Consequently, the trigonal prism is only observed in eighteen electron complexes when it is forced by the geometric constraints of the ligand. When the d_{xy}, $d_{x^2-y^2}$ set is depopulated (e.g. in a 14 electron complex) then the mixing effects no longer destabilise the system and trigonal prismatic co-ordination begins to compete with the octahedral coordination geometry. Indeed the great majority of trigonal prismatic complexes have d^2 configurations. Examples of such complexes are given in Table 8[42].

Table 8. Some examples of six co-ordinate complexes

Complex	d^n	Structure
$Mo(S_2C_2H_2)_3$	0	TP
$(Ph_4As)[Nb(S_2C_6H_4)_3]$	0	TP
$[TiF_6]^{2-}$	0	dist. OCT
$Re(S_2C_2Ph_2)_3$	1	TP
MoS_2	2	TP
OsF_6	2	OCT
$[Cr(NH_3)_6]^{3+}$	3	OCT
IrF_6	3	OCT
$Mo(CO)_2(S_2CNPr^i_2)_2$	4	dist. TP
PtF_6	4	OCT
$[Mn(H_2O)Cl_5]^{3-}$	5	OCT
$Cr(CO)_6$	6	OCT

Table 9. Some examples of five co-ordinate complexes

Complex	d^n	Structure
$[RuCl_2(PPh_3)_3]$	6	SP
$[Co(CNPh)_5]^+$	8	SP
$Fe(CO)_5$	8	TBP
$[Ni(CN)_5]^{3-}$ [a]	8	SP/dist. TBP
$[CdCl_5]^{3-}$	10	TBP
$[InCl_5]_2^-$	10	SP

[a] $[Cr(en)_3][Ni(CN)_5] \cdot 1.5\ H_2O$ has two crystallographically distinct $[Ni(CN)_5]^{3-}$ ions: one SP, the other a distorted TBP

Five co-ordination is associated with two common polyhedral geometries – the trigonal bipyramid and the square-pyramid. Both represent almost equally viable solutions to the packing and covering problems. The complementary nature of the interactions between the ligand sphere and the metal orbitals are summarised in Table 4 and examples of eighteen electron complexes based on these geometries are summarised in Table 9[43]. In common with the trigonal prism, the orbitals localised on the metal are nolonger rigorously non-bonding. For example, in the square-pyramid the d_{z^2} orbital has the same symmetry transformation properties in the C_{4v} point group as the ligand S^σ function and therefore experiences an antibonding interaction, which is mitigated by the higher lying metal s and p orbitals. Similarly, the d_{xy}, and $d_{x^2-y^2}$ orbitals in the trigonal bipyramid have the same symmetry properties as p_x, p_y (e′) and are capable of mixing with the ligand P^σ_{1s}, P^σ_{1c} functions in an antibonding fashion. The degree of antibonding character in the highest occupied orbitals will depend critically on the relative energies of the d, s and p orbitals and their respective overlaps with the ligand functions[44]. In the extreme, however, if the p orbitals are very high lying they will not be able to moderate the antibonding character in the highest occupied orbitals of the complex and the inert gas rule formalism will begin to break down. Nyholm[45] was one of the first to recognise the importance of d – p promotion energies to the inert gas formalism.

For four coordination the complementary metal d orbitals consist of the complete set of five d orbitals. The splitting of the d orbitals into t_2 and e sets is a direct consequence of the breakdown of the spherical approximation and mixing of ligand P^σ functions with the three metal d orbitals of matching t_2 symmetry. The eighteen electron configuration is associated with filling the slightly antibonding t_2 set. This can be mitigated by d – p mixing or π-bonding effects. $[Cu(MeCN)_4]^+$ and $Ni(CO)_4$ provide examples of such complexes which conform to the inert gas rule formalism[46]. An ML_N complex has S^σ, $3P^\sigma$, and $(N-4)D^\sigma$ strongly bonding M–L molecular orbitals and (9 – N) filled d orbitals localised mainly on the metal. When taken together these ligand and metal based orbitals form a complete set of functions whose nodal characteristics emulate those of an inert gas.

Adherence to the inert gas rule depends on the following factors:

(a) Complementary interactions between the ligand and central atom wavefunctions which lead to a pseudo-spherical electron distribution.

(b) Those polyhedra which give effective covering and packing on the sphere are most effective in emulating the spherical harmonic functions of a spherical shell and generating a contiguous set of S^σ, P^σ and D^σ functions.

(c) For eight, seven and six coordinate structures the complementary (9 – N)d functions are localised on the central atom and are essentially non-bonding. These orbitals can experience an additional stabilisation through π-bonding effects.

(d) For five and four coordinate structures the highest lying complementary d functions mix to some extent with the ligand P^σ functions. These antibonding interactions erode the applicability of the inert gas rule unless they are mitigated either by π-bonding effects or extensive mixing with higher lying atomic p orbitals.

Craig and Doggett have examined in some detail the radial parts of the wave functions in some metal carbonyl complexes and concluded that the electron distribution around the central atom emulates that of an inert gas. X_α calculations on $[ReH_9]^{2-}$ [33] suggest that hydrido-ligands are also able to function in this manner[25].

E. Equivalent and Localised Orbital Representations

The delocalised molecular functions described above can be transformed into a set of equivalent orbitals which are localised in specific regions following the procedures originally developed by Lennard-Jones[47]. If there is a 1:1 mapping of chemical bonds and bonding molecular orbitals then the resultant equivalent orbitals are localised in specific bond directions. For example, for methane the following molecular orbitals have been calculated using ab initio techniques (if the small contributions from the carbon 1s orbitals are ignored)[48]:

$$\psi_1(S^\sigma, a_1) = 0.58\,2s + 0.19(\sigma_1 + \sigma_2 + \sigma_3 + \sigma_4)$$

$$\psi_2(P^\sigma_0, t_2) = 0.55\,2p_z + 0.32(\sigma_1 - \sigma_2 - \sigma_3 + \sigma_4)$$

$$\psi_3(P^\sigma_{1c}, t_2) = 0.55\,2p_x + 0.32(\sigma_1 - \sigma_2 + \sigma_3 - \sigma_4)$$

$$\psi_4(P^\sigma_{1s}, t_2) = 0.55\,2p_y + 0.32(\sigma_1 + \sigma_2 - \sigma_3 - \sigma_4)$$

Four linearly independent equivalent orbitals which point towards individual hydrogen atoms can be obtained by multiplying through by the inverse of the ligand transformation matrix. The following orthogonalised and normalised functions result[49]:

$$\theta_1 = 0.29\,2s + 0.28(2p_x + 2p_y + 2p_z) + 0.58\,\sigma_1 - 0.07(\sigma_2 + \sigma_3 + \sigma_4)$$

$$\theta_2 = 0.29\,2s + 0.28(-2p_x + 2p_y - 2p_z) + 0.58\,\sigma_2 - 0.07(\sigma_1 + \sigma_3 + \sigma_4)$$

$$\theta_3 = 0.29\,2s + 0.28(2p_x - 2p_y - 2p_z) + 0.58\,\sigma_3 - 0.07(\sigma_1 + \sigma_2 + \sigma_4)$$

$$\theta_4 = 0.29\,2s + 0.28(-2p_x - 2p_y + 2p_z) + 0.58\,\sigma_4 - 0.07(\sigma_1 + \sigma_2 + \sigma_3)$$

The small (0.07) contribution from the three hydrogen atoms which are not part of the localised bond description is necessary to satisfy the orthogonality conditions. They arise because the mixing coefficients for the a_1 and t_2 molecular orbitals are not equal. They would only be equal if the 2s and 2p orbitals of carbon were degenerate and interacted equally with the hydrogen linear combinations. In those circumstances the following hybrid orbitals would form the basis of the equivalent orbital representations:

$$hy_1 = 1/2(2s + 2p_x + 2p_y + 2p_z)$$

$$hy_2 = 1/2(2s - 2p_x + 2p_y - 2p_z)$$

$$hy_3 = 1/2(2s + 2p_x - 2p_y - 2p_z)$$

$$hy_4 = 1/2(2s - 2p_x - 2p_y + 2p_z)$$

A similar procedure leads to equivalent orbital representations for the other co-ordination polyhedra described above for MH_N compounds which conform to the inert gas rule. The hybrid orbital representations approximate to those proposed by Pauling and Kimball[50,51], but because of the large differences in radial distribution functions and ionisation energies for the nd, (n + 1)s and (n + 1)p orbitals of transition metals the localisation of the equivalent orbitals will not be as complete as that described above for CH_4[52].

Nevertheless, the equivalent orbital method effectively links the molecular orbital and valence bond formalisms. The choice of equivalent orbitals is based generally on a minimisation of electron repulsion between bonds, but the analysis above suggests that the choice of ligand polyhedron is also influenced by sphere covering and packing effects.

F. Hypervalent Compounds

Although the arguments developed above have described the bonding in compounds which have either a total of eight or eighteen valence electrons, there are numerous compounds particularly of the main group elements where the inert gas rule is inapplicable. How does this come about? For the main group elements the ns and np orbitals are of similar energies and the nd orbitals are much higher lying and make only a small contribution to bonding. For species such as PF_5, SF_6, IF_7 and XeF_8^{2-} [53] ligand σ-orbitals successively provide S^σ, P^σ and D^σ functions in a manner described in detail above for transition metal compounds. Since these D^σ functions interact only weakly with the nd orbitals of the central atom they do not make a large contribution to the net bonding and the S^σ and P^σ functions provide the major bonding interactions and largely determine their stereochemistries. The s and three p functions of the central atom represent a spherically symmetric set and therefore ligand arrangements which most closely approximate to spherical are preferred, i.e. the deltahedral and closely related geometries described above. Table 1a summarizes the observed geometries for hypervalent main group molecules and demonstrates the validity of this generalisation. Because there are alternative polyhedral geometries for $N = 5, 7$ and 8 with comparable spherical characteristics many of these molecules are stereochemically non-rigid[54].

The localisation of the bonding M–L D^σ functions primarily on the ligands suggests that hypervalency is observed generally with ligands of high electronegativity. In summary, hypervalent compounds arise from the utilisation of ligand linear combinations which are matched by high lying virtual orbitals on the central atom. It is not surprising that such compounds are very much rarer for the d block transition metals, where the ligand set would be required to generate F^σ functions, i.e. adopt deltahedra with $N > 9$.

Although, the s and p orbitals of the central atom in such compounds represent a spherically symmetric situation filling of the ligand D^σ shell can lead to an asymmetric charge distribution. For example, in a trigonal bipyramid the ligand D_0^σ function is localised more extensively on the axial than the equatorial ligands, consistent with the higher negative charge associated with the axial ligand compared with the equatorial ligands. This conclusion has been confirmed by molecular orbital calculations at the ab initio level and has important ramifications for the site preferences in substituted trigonal bipyramidal compounds[55]. In octahedral compounds the ligand D_0^σ and D_{2c}^σ orbitals are occupied and the resultant electron distribution has cubic symmetry.

In hypervalent compounds the localisation of molecular orbitals utilising the equivalent orbital method is problematical because the number of strongly bonding molecular orbitals (four) is not equal to the number of bonds (five, six, seven, etc.). A common way of circumventing this orbital deficiency problem is to use combinations of two-centre two-electron and three-centre four-electron bonds as a basis for the localisation procedure. For example, in PF_5 the equatorial bonds are described in terms of localised two-centre bonds (based approximately on sp^2 hybrids) and the axial bonds in terms of a three-centre four-electron bond involving the phosphorus p_z orbital. This localisation process provides a convenient mode of rationalising the occurrence of longer $P–F_{axial}$ bonds[56].

G. Ligand Dissociation Processes

I. Spherical Co-Ordination Polyhedra

A decrease in co-ordination number can be formulated in two distinct ways: a two electron ligand, L, may either dissociate as a neutral ligand or as a dicationic species L^{2+}

leaving its lone pair of electrons on the central atom, M. Different stereochemical consequences are observed depending on whether M is a main group or transition metal atom.

In ML_N the ligand coordination polyhedron is characterised by the following linear combinations: S^σ, P^σ and $(N-4)D^\sigma$. If M is a main group atom then the S^σ and P^σ interact primarily with the ns and np valence orbitals of the central atom and the D^σ only weakly with nd. The resultant m.o.s are occupied by 2 N electrons, 8 in S^σ and P^σ and $2(N-4)$ in D^σ. A decrease in the co-ordination number resulting from the dissociation of L leads to a decrease in the number of D^σ functions required to define the ligand polyhedron. The new ligand environment rearranges to form the $N-1$ deltahedron with S^σ, P^σ and $(N-5)D^\sigma$ ligand functions, which on interaction with the central atom orbitals leads to $(N-1)$ stable molecular orbitals. This smooth transformation to a second deltahedron with one fewer vertices is possible because the D^σ functions are localised almost exclusively on the ligand atoms.

If M is a transition metal the eighteen electron configuration is achieved for the following sets of matching ligand and metal based functions:

$$\underbrace{S^\sigma,\ P^\sigma,\ (N-4)D^\sigma,}_{\text{localised mainly on ligand sphere}} \qquad \underbrace{(9-N)d}_{\text{localised mainly on metal}}$$

A decrease in the co-ordination number leads to a complementary set of orbitals if the ligand D^σ function lost as a result of the decrease in co-ordination number is replaced by a filled metal d orbital with quantum numbers identical to those for the lost D^σ function. These conditions are met only if the process leading to the $(n-1)$ deltahedron results from the dissociation of L^{2+}. Therefore, ligand dissociation processes from main group and transition metal compounds have the following complementary stereochemical consequences.

$$ML^{2+}_{N-1} \xleftarrow{\ -L^{2+}\ } ML_N \xrightarrow{\ -L\ } ML_{N-1}$$

M = transition metal atom		M = main group atom
Deltahedron	*Deltahedron*	*Deltahedron*

The complementary nature of these processes arises because for transition metals the lost D^σ function is replaced by a matching filled d orbital. The lost D^σ function does not require replacement for main group compounds because it is localised exclusively on the ligand sphere.

Interestingly the alternative dissociative processes, i.e.

$$ML^{2+}_{n-1} \xleftarrow{\ -L^{2+}\ } ML_N \xrightarrow{\ -L\ } ML_{N-1}$$

M = main group atom	M = transition metal atom

also take on a complementary form, but in this instance it is not generally possible to maintain the deltahedral geometries.

The loss of L^{2+} from a main group compound, e.g. $SF_6 \rightarrow BrF_5$; $BrF_5 \rightarrow XeF_4$ results in the loss of a ligand linear combination, and an additional electron pair has to be accommodated in the molecular orbitals which remain. In *closo*-ML_N S^σ, P^σ and $(N-4)D^\sigma$ m.o.'s are occupied by 2N electrons, if the loss of L^{2+} leads to S^σ, P^σ and $(N-5)D^\sigma$ m.o.'s then the remaining 2N electrons exceed those required for filling all bonding m.o.'s by 2, and would have to occupy antibonding m.o.'s if the ML_{N-1} polyhedron was deltahedral.

Similarly loss of L from an 18 electron transition metal complex leads to a hole in the d manifold:

$$S^\sigma P^\sigma (N-4) D^\sigma \ (9-N)d \rightarrow S^\sigma P^\sigma (N-5) D^\sigma \ (9-N)d$$

The (N−1) complex no-longer has a complementary set of filled ligand and metal orbitals and therefore may not necessarily adopt a deltahedral geometry.

The bonding situations are related in the two cases. For the main group ML_{N-1} compound the antibonding character of the additional filled m.o. must be minimised and for the corresponding transition metal complex the energy of the hole in the d manifold must be raised as much as possible. This increases the h.o.m.o.-l.u.m.o. gap and hence maximises the bonding with the remaining ligands. Both these effects are achieved by adopting non-spherical polyhedra. Therefore, it is necessary to digress a while to consider how such polyhedra may be defined as fragments of deltahedra.

II. Non-Spherical Co-Ordination Polyhedra

The definition of the linear combinations of atomic orbitals of the ligands in ML_N given in Eq. IV is a reasonable approximation for spherical polyhedra particularly when N is large. Many compounds, however, have co-ordination polyhedra which are distinctly non-spherical. Some examples are illustrated below:

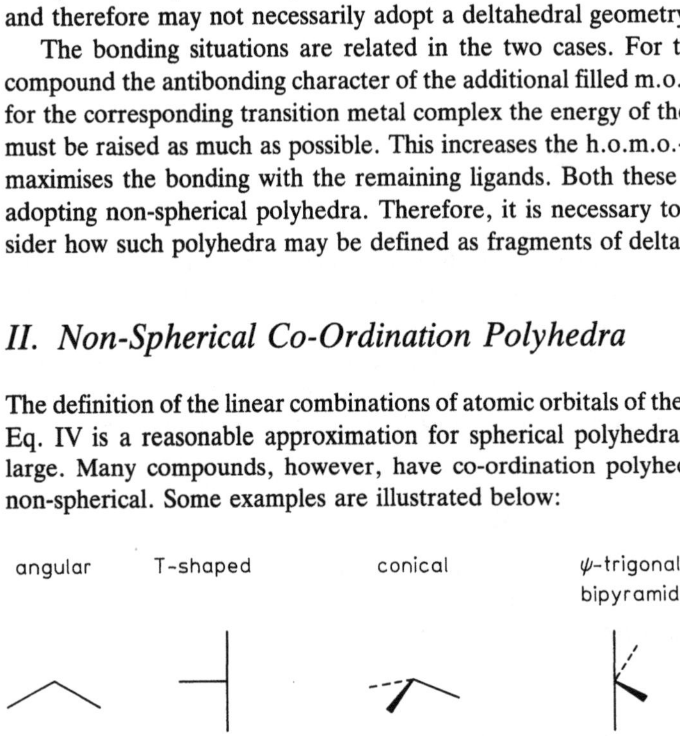

These polyhedra are fragments of *closo*-polyhedra, and following the precedent set for borane anions they can be described as *nido*-, *arachno*- and *hypho*- if they can be represented as fragments of deltahedra with one, two and three vertices missing respectively[58]. For these fragments it is useful sometimes to make a topological distinction between planar and non-planar polyhedra.

The grossly non-spherical nature of these fragments means that their linear combinations are not well represented as spherical harmonic expansions. They can, however, be described as subsets of the parent polyhedron and their linear combinations can be described in those terms. The arguments are similar to those developed above for capped polyhedra. For example, the linear combinations of a square-pyramid can be derived

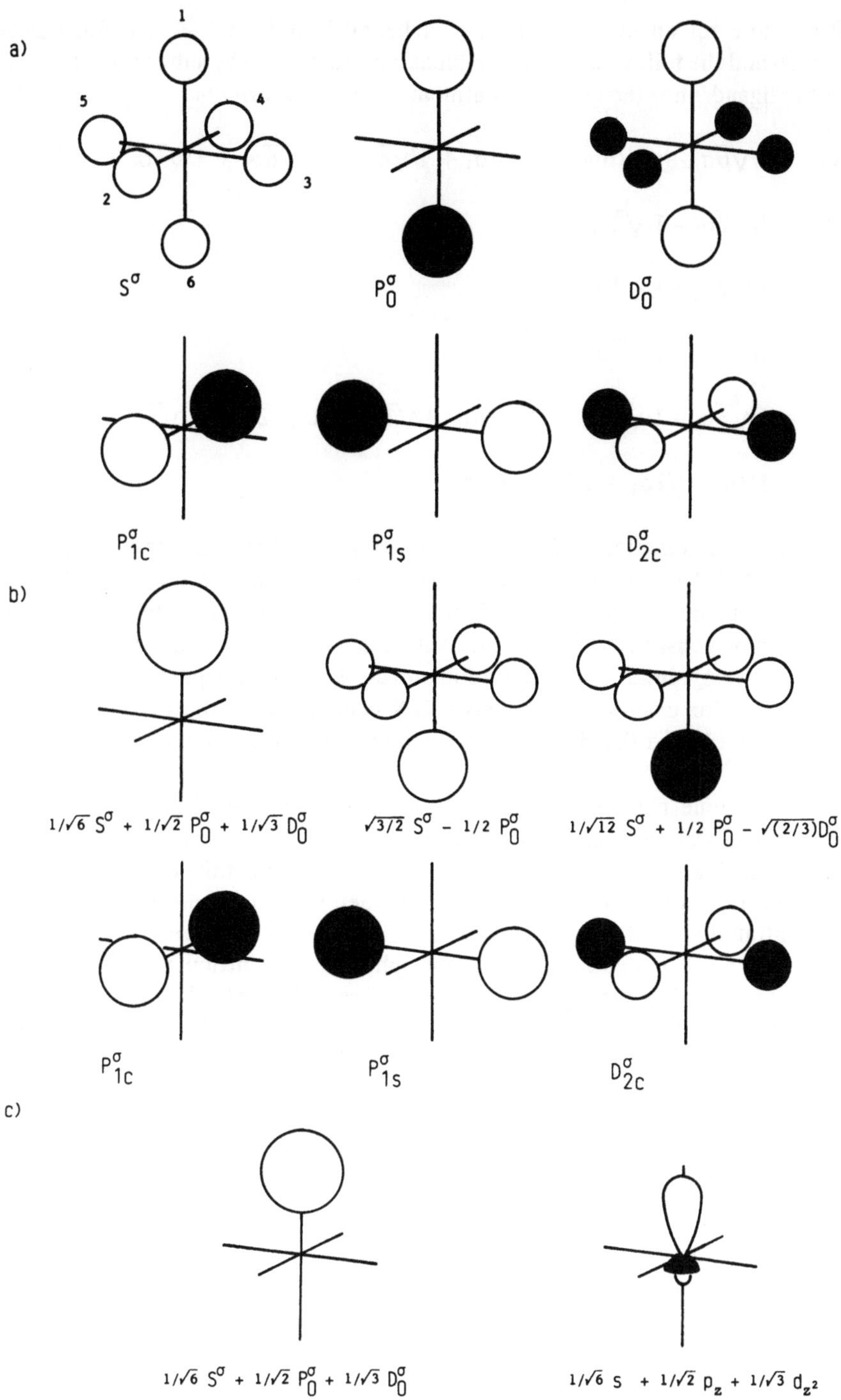

Fig. 20. (**a**) Illustrates the linear combinations of ligand orbitals in an octahedral compound. In (**b**) linear combinations of those shown in (a) can result in localisation at position 1. (**c**) represents schematically the relationship between the localised linear combination and an equivalent orbital on the central atom

from those for an octahedron. An octahedral ligand set 1–6 is defined as shown in Fig. 20 and the following linear combinations are derived by subsituting the co-ordinates of the ligands into the spherical harmonic expansion formulae.

$$S_0^\sigma = 1/\sqrt{6}\sigma_1 + 1/\sqrt{6}\sigma_2 + 1/\sqrt{6}\sigma_3 + 1/\sqrt{6}\sigma_4 + 1/\sqrt{6}\sigma_5 + 1/\sqrt{6}\sigma_6$$

$$P_0^\sigma = 1/\sqrt{2}\sigma_1 - 1/\sqrt{2}\sigma_6$$

$$P_{1c}^\sigma = 1/\sqrt{2}\sigma_2 - 1/\sqrt{2}\sigma_4$$

$$P_{1s}^\sigma = 1/\sqrt{2}\sigma_3 - 1/\sqrt{2}\sigma_5$$

$$D_0^\sigma = 1/\sqrt{3}\sigma_1 - 1/\sqrt{12}\sigma_2 - 1/\sqrt{12}\sigma_3 - 1/\sqrt{12}\sigma_4 - 1/\sqrt{12}\sigma_5 + 1/\sqrt{3}\sigma_6$$

$$D_{2c}^\sigma = 1/2\sigma_2 - 1/2\sigma_3 + 1/2\sigma_4 - 1/2\sigma_5$$

The linear combination of the octahedral symmetry adapted linear combinations $1/\sqrt{6}S^\sigma + 1/\sqrt{2}P_0^\sigma + 1/\sqrt{3}D_0^\sigma$ is localised exclusively at position 1 on the sphere (see Fig. 20b) and the remaining five linear combinations are delocalised over the other five locations and may be constructed to be equivalent to the symmetry adapted linear combinations for a square-pyramid. This localisation process is useful for a perturbation analysis of the bonding in for example octahedral ML_5X complexes. In the limit of X being a very weakly held ligand the situation clearly approximates to a square-pyramidal ML_5 complex[59)].

Ligand lone pair orbitals represent the most obvious of the alternative modes of generating localised regions of electron density on the surface of the ligand sphere. A linear combination of atomic orbitals on the central atom of the type $as + bp_z + cd_{z^2}$ has a localised region of electron density which when projected onto to the ligand sphere is concentrated around the polar region (see Fig. 20c). Therefore, it is possible for the N regions of electron density on the ligand sphere to be contributed in part by ligand lone pair orbitals and in part by a combination of central atom orbitals.

The linear combination $as + bp_z + cd_{z^2}$ produces a region of electron density in the polar region and in addition will be orthogonal to the symmetry adapted ligand linear combinations of the *nido*-polyhedron if the coefficients a, b and c are equal to those calculated for the localised orbital around position 1, i.e. in the case of an octahedron $1/\sqrt{6}S^\sigma + 1/\sqrt{2}P_0^\sigma + 1/\sqrt{3}D_0^\sigma$ gives $a = 1/\sqrt{6}$, $b = 1/\sqrt{2}$ and $c = 1/\sqrt{3}$. This orthogonality relationship means that in an ideal situation where the hybrid orbital approximation is valid, an electron pair occupying $1/\sqrt{6}s + 1/\sqrt{2}p_z + 1/\sqrt{3}d_{z^2}$ on the central atom will be essentially non-bonding with respect to the bonding molecular orbitals associated with the square-pyramidal coordination polyhedron[60)].

We have an additional complementary principle; a *closo*-polyhedron may be emulated by a *nido*-set of ligand lone pair orbitals and an orthogonal combination of central atom orbitals. This principle could equally well have been derived from the equivalent orbital method described in Sect. D since the equivalent orbitals are orthogonal and localised in specific bond directions. Analogous complementary sets of ligand and atomic hybrid orbitals can be derived for *arachno*- and *hypho*-polyhedra. Consequently the structures of idealised main group compounds of the general type ML_N, $ML_{N-1}E$, . . ,

Table 10. Residual S^σ, P^σ and D^σ character on the fragments of the ligand sphere in *nido*- and *arachno*-coordination polyhedra

N	Closo			Nido			Arachno			Planar		
	S^σ	P^σ	D^σ	S^σ	P^σ	D^σ	S^σ	P^σ	D^σ	S^σ	P^σ	D^σ
3	1.00	2.00	-	0.75	2.25	-	[a]0.63 [b]0.59	1.67 1.83	0.71 0.57	0.56	2.00	0.44
4	1.00	3.00	-	[c]0.78 [d]0.81	2.50 2.33	0.72 0.85	0.67	2.00	1.33	0.67	2.00	1.33
5	1.00	3.00	1.00	0.83	2.50	1.67	[a]0.69 [b]0.73	2.20 2.10	2.11 2.17	0.78	2.00	2.22
6	1.00	3.00	2.00	[c]0.89 [d]0.84	2.50 2.60	2.61 2.56						
7	1.00	3.00	3.00									

[a]Di-equatorially localised
[b]Mono-axially, mono-equatorially di-localised
[c]Axially mono-localised
[d]Equatorially mono-localised

where E represents the number of electron pairs in excess of those required for M–L bonding, are based on the *closo*-polyhedron favoured by ML_N. $ML_{N-1}E$ adopts a *nido*-structure and $ML_{N-2}E_2$ an *arachno*-structure. Examples, of main group compounds which show these structural inter-relationships are commonplace, and some are given in Table 1 a.

In some instances, e.g. a trigonal bipyramid and a pentagonal bipyramid, there exists a choice between axial and equatorial vertices for the *closo*- to *nido*-transformation. It is a simple matter to calculate the percentage S^σ, P^σ and D^σ character associated with localising the ligand functions either on the axial or equatorial sites and the procedure is given in the Appendix. The results are given in Table 10 and summarised below for the trigonal bipyramid:

The localisation process reveals more total D^σ character on the axial than the equatorial site. If the central atom had degenerate s, p and d atomic orbitals which overlapped equally with the ligand orbitals then they could provide with equal facility a complementary hybrid orbital at either position. For main group atoms the ns and np valence orbitals are substantially more stable than the nd orbitals. Therefore, it is energetically more favourable to utilise a *nido*-ligand polyhedral fragment with the maximum amount of D^σ orbital character. This will naturally generate an orbital localised mainly on the central atom which has the maximum amount of s and p character and will accomodate two electrons. The complementary nature of these preferences is summarised schematically

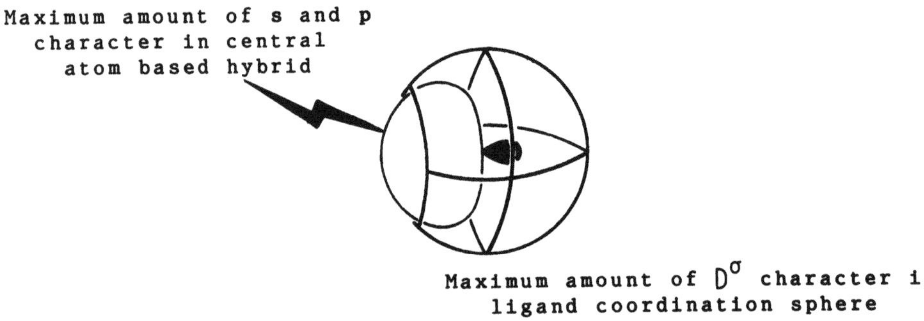

Fig. 21. The complementary nature of the ligand fragment and the equivalent orbital on the central atom. When taken together they generate an approximately spherical distribution of electron density

in Fig. 21. For the trigonal bipyramid removal of an equatorial ligand has the effect of retaining the maximum amount of D^σ character on the remaining fragment of the ligand sphere. The observed geometries of PF_5, SF_4, ClF_3 and XeF_2 can be rationalised by this principle. It is noteworthy that this explanation is at variance with that commonly given in elementary textbooks and based on the valence shell electron pair repulsion theory. The geometries of molecules such as SF_4, ClF_3 and XeF_2 are rationalised in terms of minimising electron pair repulsions[61)] rather than the distribution of s, p and d character between the ligand sphere and the central atom.

The occurrence of BrF_5 and XeF_4 based on the parent octahedral geometry observed in SF_6 can similarly be rationalised by this principle. For XeF_4 the choice between putting the two vacancies in cis- or trans-locations of the octahedron cannot be made on the basis of the relative percentages of d, s and p character on the ligand sphere (see Table 10). Nevertheless, given that two electron pairs are localised primarily on the Xe atom and have a high percentage of s and p character the planar arrangement of fluorines can take advantage of the nodal plance of the filled p orbital and minimise the antibonding interactions. The complementary nature of the ligand and central atom electron distributions are represented schematically below.

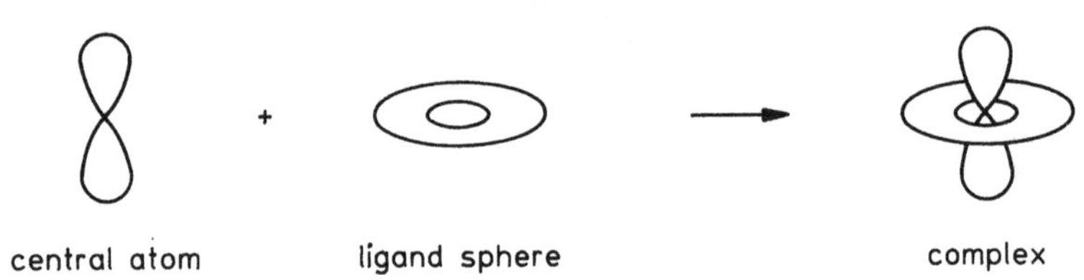

Planar ligand arrangements are observed for all ML_3E_2 molecules (e.g. OF_2 and ClF_3) for the same reason. When the molecule has three electron pairs in equivalent orbitals localised on the central atom then they will have a high percentage of s, p_x and p_y character and the ligands lie on the nodal line of the $p_x^2 + p_y^2$ torus of electron density. such a linear arrangement is found in XeF_2.

The discussion above has demontrated how alternative *nido-* and *arachno-*fragments lead to different amounts of D^σ character on the ligand part of the sphere. The concept of maximising the amount of D^σ character on the ligand sphere can also be extended to provide a more detailed account of the angular geometries of main group molecules, since it will be shown that distortion away from the idealised geometries influence the relative amounts of S^σ, P^σ and D^σ character. For molecules where the central atom does not have d valence orbitals then angular changes can influence the relative amounts of S^σ and P^σ character on the ligand sphere.

III. Bond Angles in Nido- and Arachno-MH_N

The ligand atoms in a main group compound distort away from the idealised geometry. The data summarised in Table 11 demonstrate that the distortion always occurs in the same sense, i.e. away from the missing vertices of *nido-* and *arachno-*polyhedra[61, 62].

In the tetrahedrally based series CH_4, NH_3 and OH_2 the linear combinations of hydrogen is orbitals are based exclusively on S^σ and P^σ spherical harmonic expansions. Since the 2s orbitals are lower lying than the 2p orbitals the S^σ function in CH_4 has more carbon character than the P^σ functions (see Sect. D.). In an idealised *nido*-MH_3 tetrahedral fragment, with H–M–H = 109.45°, the central atom would ideally have an equivalent orbital pointing away from the hydrogen atoms with coefficients $1/\sqrt{2}\,s + \sqrt{3}/2\,p_z$. However, since the 2s orbital is more stable than the 2p orbital for all main group atoms a

Table 11. Summary of angular distortions from the idealised values in main group co-ordination compounds

	CH_4 109.5°			
	NH_3 106.8°	PH_3 93.3°	AsH_3 91.8°	
	OH_2 104.5°	SH_2 92.2°	SeH_2 91°	
		ClF_3 87.5°	BrF_3 86.2°	
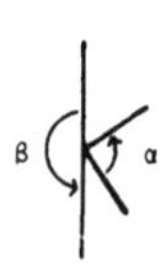		SF_4 α 101° β 187°		
			BrF_5 84°	$[XeF_5]^+$ 79°

more stable structure is achieved when the central atom contributes more s character to the equivalent orbital. The balance in S and P character in the molecule is maintained if the linear combinations of the hydrogen atoms have more P^σ character. This redistribution of S and P character between the central atom and the ligand sphere can only be achieved if the hydrogen atoms distort away from the idealised tetrahedral positions.

The effects of varying the % p_z character on the the nodal cone of an as + bp_z hybrid orbital are shown in Fig. 22[63]. As the s character is increased the nodal planes move away from the vacant co-ordination site leading to a decrease in the H–M–H bond angle. The movement of hydrogen atoms in these directions maintains the orthogonality relationship and results in a corresponding increase in ligand P^σ character. Table 11 gives examples of pyramidal MX_3 molecules where the X–M–X angle is always less than the ideal tetrahedral angle. Analogous distortions are observed in other *nido*- and *arachno*-MX_NE_M molecules, and the results are summarised in the Table.

For hypervalent compounds such as BrF_5 and SF_4 the d orbitals cannot make a contribution to the hybrid orbital which is equal to that predicted on purely geometric grounds. Consequently the hybrid has an excess of s and p character. The deficiency of d character has the effect of changing the positions of the nodal cone in the manner

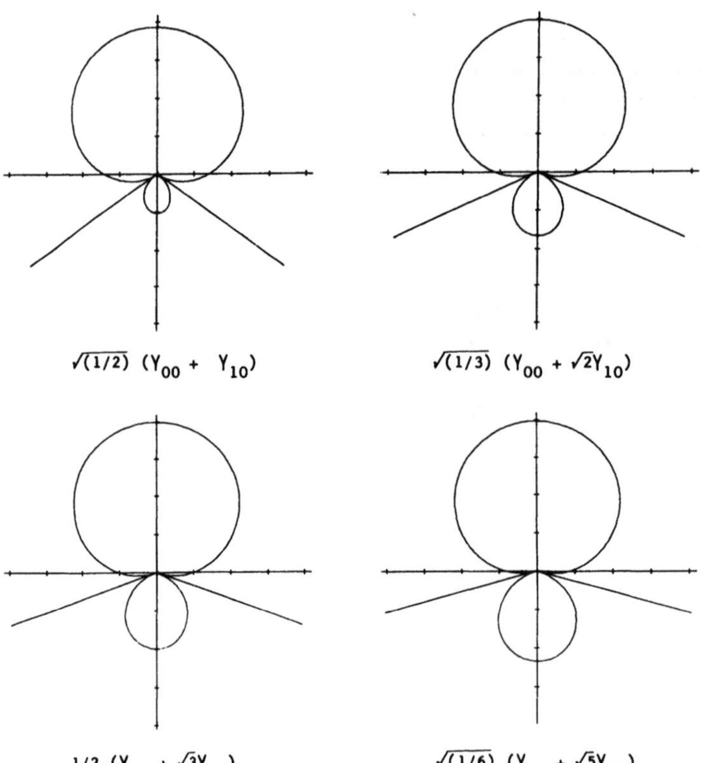

Fig. 22. Polar plots of linear combinations of the Y_{00} (s) and Y_{10} (p_z) spherical harmonics in the xz plane. Particularly noteworthy is the way in which the angle between the nodal lines increases as the contribution from Y_{10} increases. In molecular terms an increase in s character in the equivalent orbital localised on the central atom leads to a decrease in the X–M–X bond angle

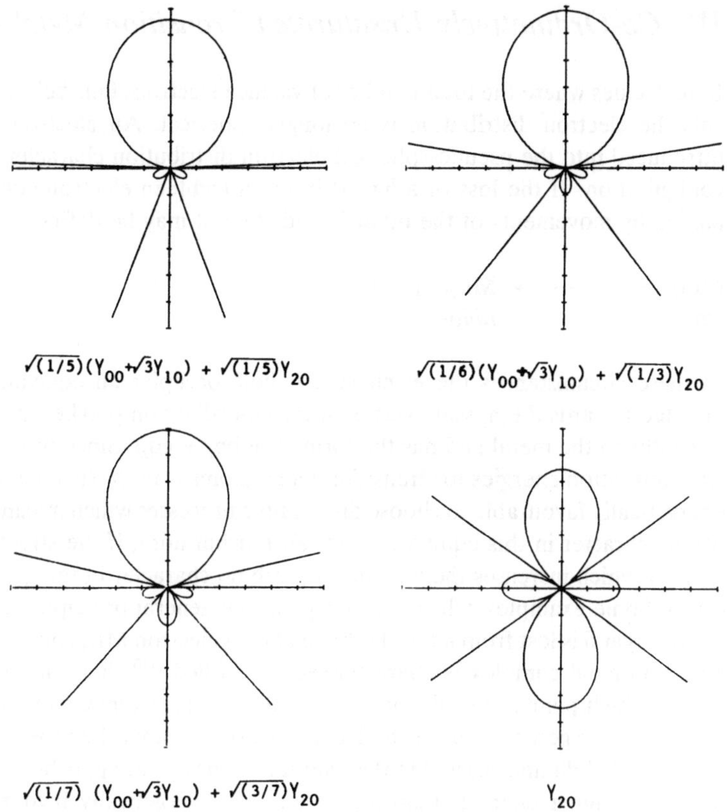

Fig. 23. Polar plots of linear combinations of the Y_{00} (s) Y_{10} (p_z) and Y_{20} (d_{z^2}) spherical harmonics. The $Y_{00}:Y_{10}$ ratio is kept constant and the figure illustrates the influence of increasing the Y_{20} character. As the percentage of Y_{20} (d_{z^2}) is decreased the nodal lines move away from the direction of maximum electron density

illustrated in Fig. 23. Distortion of the ligand away from the idealised position places them along the nodal cones and consequently preserves the orthogonality between the ligand-central atom molecular orbitals and the hybrid orbital.

The arguments developed here are of course similar to those developed by Bent and used to rationalise not only geometric features, but also spectroscopic parameters such as coupling constants[64]. It is also noteworthy that Parr[15] has calculated with great accuracy the bond angles in a wide range of main group molecules based on the electrostatic interactions between electron density in a hybrid orbital and point charges in the ligand position. Hall has also noted the important stereoactive role of the s orbital on the central atom[10].

In summary those main group molecules which have structures based on fragments of *closo*-polyhedra can be described in terms of complementary locations of electron density on the surface of a sphere, which when taken together approximate to those of the parent coordination polyhedron. The unequal contribution of the valence s, p and d orbitals to the electron density localised at the missing position(s) result in the observed angular distortions away from the idealised *nido*- and *arachno*-geometries.

IV. Co-Ordinatively Unsaturated Transition Metal Complexes

In molecules where the total number of valence electrons falls below that of the inert gas rule the electron distribution is no longer spherical. An electron pair hole has been introduced into the pseudo-spherical electron distribution characteristic of the inert gas configuration. If the loss of a ligand from an eighteen electron complex is not accompanied by movements of the other ligands then it may be defined as follows[65]:

$$\underset{closo\text{-}}{ML_N} \longrightarrow \underset{nido\text{-}}{ML_{N-1}} + L$$

In these circumstances the electron pair hole occupies an equivalent orbital which is directed towards the missing vertex of the co-ordination polyhedron, is localised predominantly on the metal and has the form as + bp$_z$ + cd$_{z^2}$. Since the trends in the valence state ionisation energies for transition metal atoms is nd > (n + 1)s > (n + 1)p then it is energetically favourable to choose dissociative processes which maximise the amount of p orbital chracter in this equivalent orbital. Furthermore, if the structure can distort in a manner which increases the amount of p orbital character in the quivalent orbital it will do so. Some examples will serve to explain the general principles involved.

If a ligand is lost from a tetrahedral eighteen electron ML_4 complex then the resultant 16e pyramidal complex is characterised by a filled d^{10} shell and an empty equivalent orbital which points towards the vacant site and approximates to an sp^3 orbital, $1/2\,s + \sqrt{3}/2\,p_z$. If the geometry of the molecule distorts to planar then the amount of s character in this orbital diminishes and in the limit it becomes a pure p_z orbital. Since p orbitals are always at higher energies than the s then it is energetically favourable to localise the orbital character in this fashion. Loss of a second ligand to generate a 14e ML_2 complex generates an equivalent orbital which approximates to sp^2 if an angular geometry is

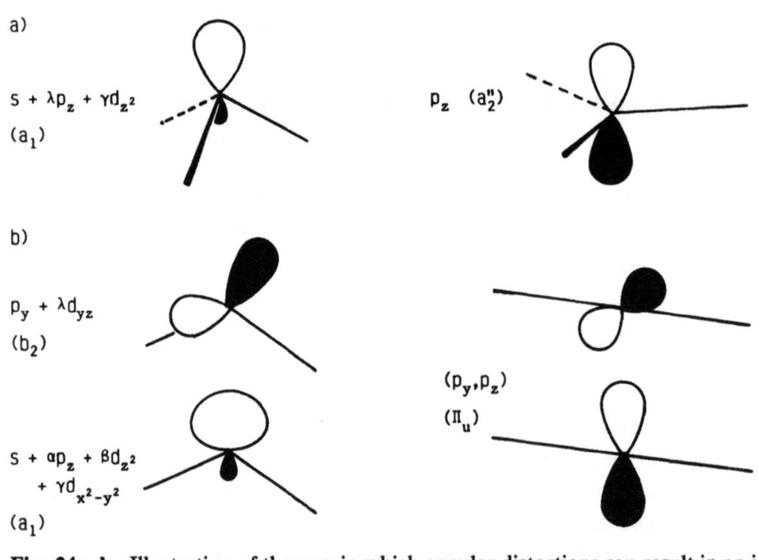

Fig. 24a, b. Illustration of the way in which angular distortions can result in an increase in p orbital character. In (**a**) from conical to planar and in (**b**) from angular to linear

Table 12. Examples of 16 electron and 14 electron complexes of the B metals

Complex	n_e	Structure
$[Ag(NH_3)_2]^+$ Au(CN)(CNMe) $HgCl_2$ HgMe(CN) $HgMe_2$	14	linear
$[SMe_3][HgI_3]$ $InMe_3$	16	trigonal planar

retained. For a linear L–M–L moiety two electron pair holes in orthogonal p orbitals are generated. These processes are represented schematically in Fig. 24. Trigonal planar 16 electron and linear 14 electron complexes are particularly prevalent for the heavier Group I b and II b metals and some examples are given in Table 12. Similar arguments apply to main group molecules which are co-ordinatively unsaturated, e.g. BeF_2 – linear and BF_3 – trigonal planar. The importance of s-p promotion energies in complexes which do not conform to the eighteen electron rule was first noted by Nyholm[45].

Five co-ordinate d^8 eighteen electron complexes are based either on the square-pyramid or the trigonal bipyramid. Their pseudo-spherical electron distributions arise from the following matching sets of ligand and metal spherical harmonic wave functions (see also Table 4):

square pyramid:	S^σ, P^σ, D^σ_{2c}	d_{z^2}, d_{xz}, d_{yz}, d_{xy}
trigonal bipyramid:	S^σ, P^σ, D^σ_0	d_{xz}, d_{yz}, d_{xy}, $d_{x^2-y^2}$
	Ligand	Metal

The loss of a ligand from either a basal site in the square-pyramid or one of the sites of the trigonal bipyramid results in the creation of an electron pair hole in an equivalent orbital with s, p and d character[66]. In contrast the loss of the apical ligand from the square-pyramid generates a square-planar complex with a p_z orbital perpendicular to the ligand plane which can accommodate the electron pair hole. These processes are represented schematically in Fig. 25. Therefore sixteen electron complexes of the platinum metals show such a strong stereochemical preference for square-planar geometries, particularly when the atomic s-p promotion energy is large.

Loss of a second ligand from the square pyramid to generate a fourteen electron ML_3 complex cannot lead to the formation of electron pair holes in orthogonal p orbitals, which is possible only for linear complexes. Consequently the second electron pair hole must be accomodated in an equivalent orbital directed at one of the vertices of the square plane (see Fig. 26). This orbital approximates to a dsp^2 hybrid orbital, which is considerably lower lying than p_z by virtue of the increased % of s and d character. Consequently such complexes are far more reactive and only a single example, viz $[Rh(PPh_3)_3]^+$, of a fourteen electron T-shaped complex has been structurally characterised. It is only stable in the presence of solvents with low nucleophilicities[67, 68].

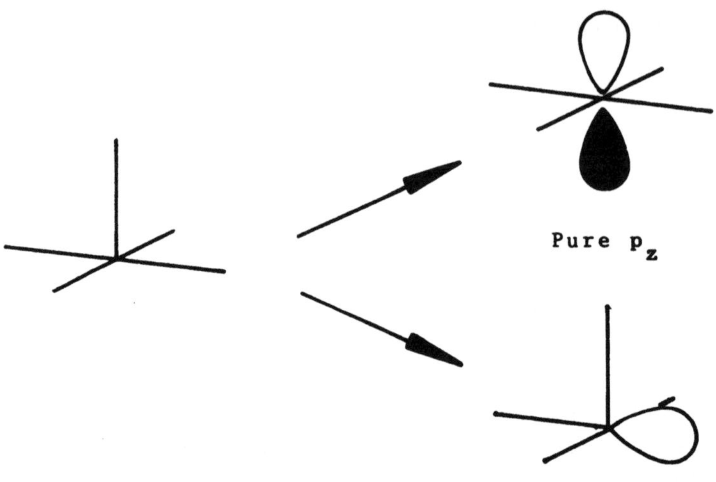

Fig. 25. Illustrations of ligand loss from a square-pyramidal 18 electron complex. Loss of an axial ligand results in an electron pair hole in a p_z orbital whereas loss of an equatorial ligand leads to a s, p, d equivalent orbital. The former is the more favourable energetically because the orbital energies for the central atom are nd < (n + 1)s < (n + 1)p

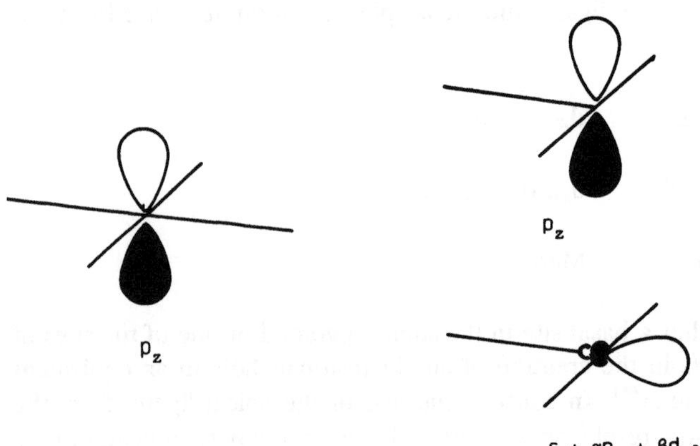

Fig. 26. Illustration of ligand loss from a 16 electron square-planar complex, which results in an out-pointing $s + p_x + d_{x^2}$ equivalent orbital

Octahedral eighteen electron complexes are extremely common and ligand loss is a key step in their thermal and photochemical reactions. The loss of a ligand results in the location of an electron pair hole in an equivalent orbital which approximates to a d^2sp^3 hybrid. In principle the formation of a d^6 pentagonal planar complex could result in the localisation of this orbital into a p_z orbital, but this is sterically unfavourable and the square-pyramidal geometry is therefore favoured. It is noteworthy that in a square pyramidal ML_5 complex the percentage p orbital character of the vacant hybrid orbital is increased by bending the ligands away from the missing vertex (see Fig. 23) yet the observed distortion usually occurs in the opposite sense. π-bonding effects are probably responsible for the observed effects. Some examples, of square pyramidal d^6 complexes are given in Table 9. The metal carbonyl complexes $Mo(CO)_5$ and $W(CO)_5$ are particularly susceptible to nucleophilic addition and co-ordinate even methane and xenon weakly[69–71].

The loss of a second and third ligand to form $M(CO)_4$ and $M(CO)_3$ also results in geometries based on the octahedron (see Fig. 27). In these examples, the electron pair holes occupy cis-sites of the octahedron. This stereochemical preference can be rationalised also in terms of maximisation of p orbital character in the vacant orbitals. Consider trans-$M(CO)_4$: the vacancies share a common p orbital and the electron pair holes are located in p_z and d_{z^2}. Whereas in cis-$Mo(CO)_4$ the vacancies share a pair of p orbitals and the electron pair holes are located in a pair of equivalent orbitals with substantial p_x and p_y character.

The generation of fragments based on the octahedron for co-ordinatively unsaturated $M(CO)_n$ fragments is central to the *isolobal* concept introduced by Hoffmann and Min-

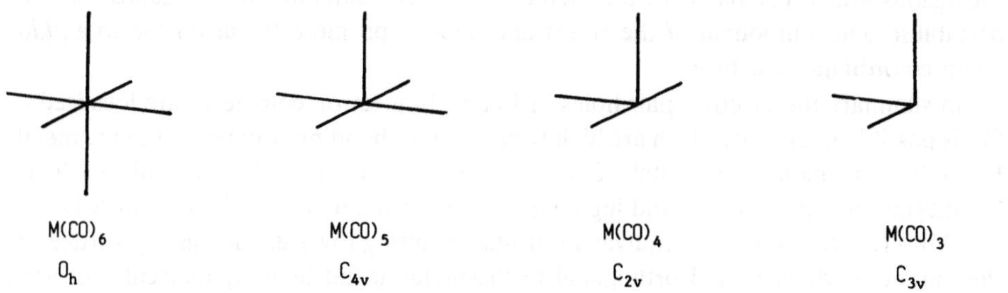

Fig. 27. Summary of stereochemical preferences for d^6 metal carbonyl fragments. Each is based on the octahedron and the vacancies adopt mutually cis-locations.

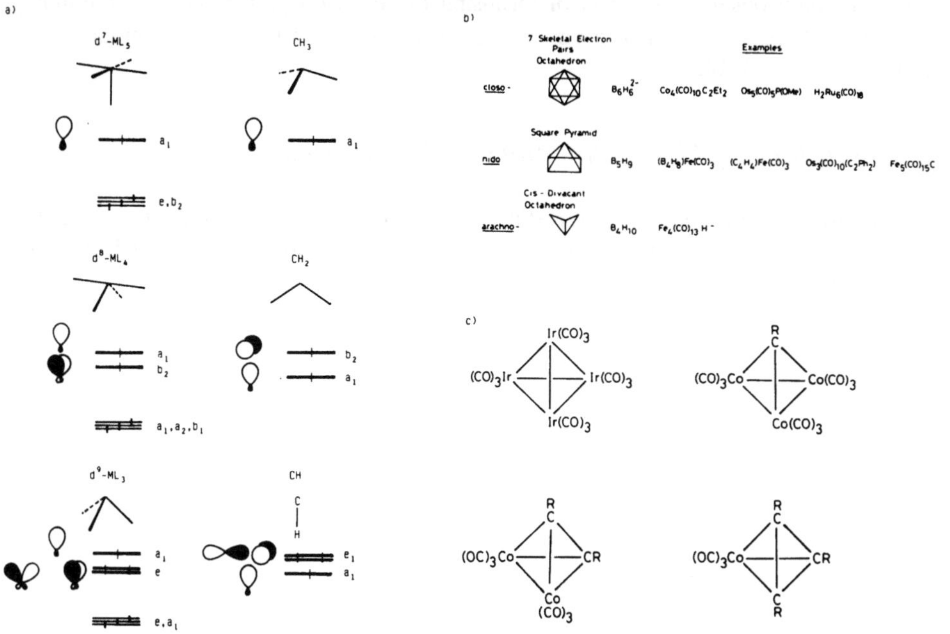

Fig. 28. (**a**) illustration of the isolobal analogy between $M(CO)_n$ and C–H fragments, (**b**) examples of isolobal replacements of the B–H and C–H fragments by $M(CO)_3$ fragments, and (**c**) successive replacement of the $M(CO)_3$ fragments by C–R in the tetrahedral $[M(CO)_3]_4$ cluster. See Refs. 73 and 74 for more detailed analyses of its use

gos[72] and disccussed at some length in Hoffmann's Nobel Lecture[73]. The *isolobal* connection between main group and transition metal fragments is illustrated in Fig. 28 together with some examples of its application[74].

For seven, eight and nine coordinate complexes the loss of a ligand can create an equivalent orbital vacancy in much the same way. The percentage p character associated with this orbital diminishes as the co-ordination number increases. Therefore, the creation of a vacancy in a d orbital which is orthogonal to the bonding metal-ligand molecular orbitals begins to compete with this alternative. For example, it is common to find d^0, d^2 and d^4 octahedral complexes with electron pair holes in the non-bonding t_{2g} orbitals; also common are d^0 and d^2 seven coordinate complexes and d^0 eight coordinate complexes. Examples of these co-ordinatively unsaturated complexes are given in Table 8[75]. Some examples of *nido-* seven co-ordinate d^4 complexes are observed when π-donor and acceptor ligands are co-ordinated to the metal[76–78]. These combinations of ligands serve to destabilise one component of the t_{2g} set and thereby promote the distortion to a *nido-* seven co-ordinate structure.

In summary the electron pair holes in 16 and 14 electron complexes are localised as far as possible in orbitals which are high lying and non-bonding with respect to the metal-ligand bonding molecular orbitals. For ML_4 (d^8), ML_3 (d^{10}) and ML_2 (d^{10}) this is achived by utilising the non-bonding and high lying nature of the metal p orbitals. For ML_3 (d^8) and ML_N (d^6) (N = 3, 4, 5) equivalent orbitals pointing towards the missing vertices of the parent polyhedron and orthogonal to the metal-ligand bonding molecular orbitals provide the most effective mode of localising the electron pair hole. For the earlier transition metals electron pair holes in the non-bonding d orbitals provide an alternative and commonly observed method of localising the electron pair holes in non-bonding orbitals localised on the central atom. These conclusions are represented schematically in a matrix form in Table 1 b.

V. Summary of Complementary Relationships

In the previous sections an interesting complementary relationship governing the stereochemistries of main group and transition metal compounds has unfolded. It can be economically represented by the following scheme:

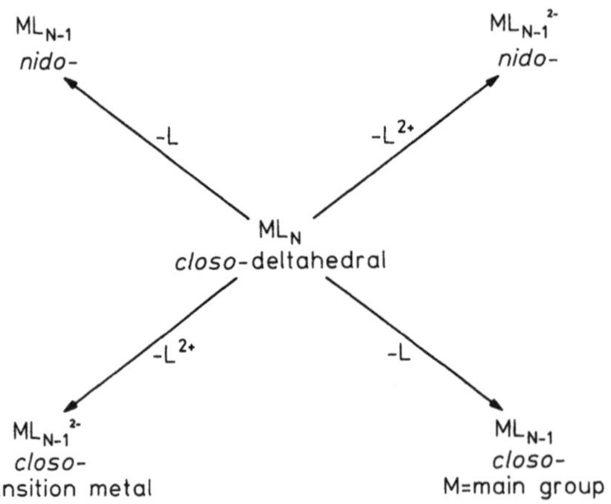

The relationships can be illustrated clearly by reference to a couple of specific examples. For the octahedral compounds $Mo(CO)_6$ and SF_6 the geometric changes accompanying the loss of L or L^{2+} take the following form:

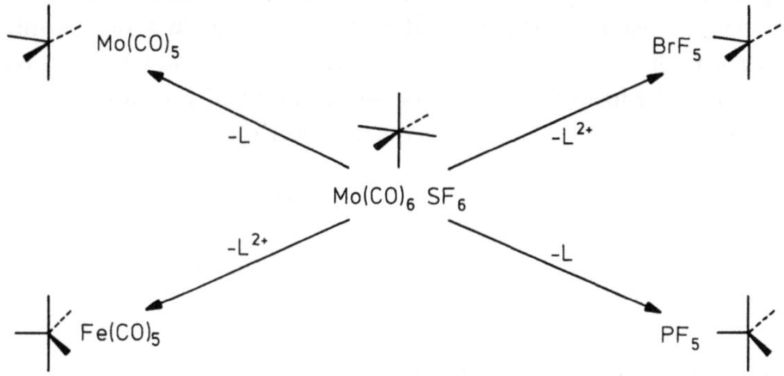

The same structures are generated on the right and left hand sides of the figure because for the octahedron all vertices are equivalent and for six co-ordination there is not an alternative co-ordination geometry. The situation is somewhat more complex for five coordinate structures where there exist two polyhedral co-ordination geometries and symmetry distinct vertices. The results are summarised below for $Ni(CN)_5^{3-}$ and PF_5.

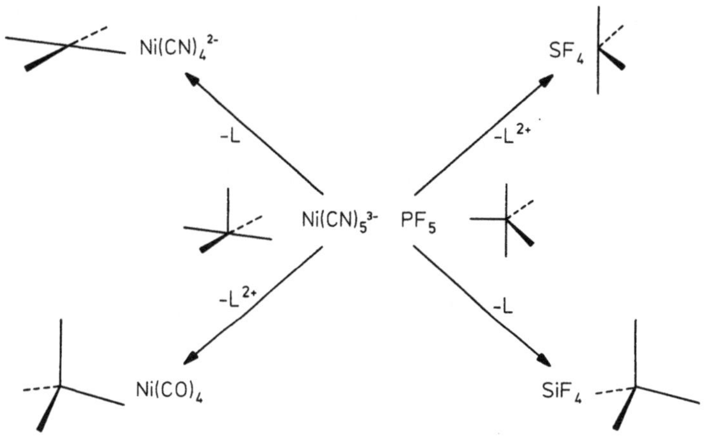

The complementary relationship has its origins in the different atomic orbital energy level orderings, i.e. ns > np ≫ nd for main group atoms and nd > (n + 1)s ≫ (n + 1)p for transition metal atoms. If M is a transition metal and ML_N an 18 electron complex (N > 4) then the electron pair holes in the d manifold of orbitals are matched by bonding molecular orbitals localised primarily on the ligands, e.g.

Trigonal bipyramid d_{z^2} matched by filled D_0^σ

Octahedron d_{z^2}, $d_{x^2-y^2}$ matched by filled D_0^σ, D_{2c}^σ.

Filling one of the electron pair holes in the d manifold requires the loss of L^{2+} and the adoption of the deltahedral geometry characteristic for N−1, i.e.

Trigonal bipyramid ⟶ tetrahedron
Octahedron ⟶ trigonal bipyramid

For the corresponding main group hypervalent compounds a similar situation pertains except that the d functions localised on the central atom are now high lying and vacant. There are (9−N)d orbitals on the central atom which are not used in σ-bonding. The loss of a ligand is accompanied by the loss of a D^{σ} function and consequently an additional d orbital which is not σ-bonding is generated as long as the new polyhedral geometry is also *closo-*.

H. π-Bonding Effects

I. General Mathematical Considerations

In Sect. D the importance of π-bonding effects in the development of the inert gas rule formalism was stressed but not discussed in detail. In ML_N ($N < 9$), where M is a transition metal atom, there are filled orbitals largely localised on the central atom, which are non-bonding or slightly anti-bonding with respect to the ligand σ-orbitals. The manner in which π-acceptor ligands stabilise these orbitals is discussed in Sect. H. II and π-donor ligands destabilise them is discussed in Sect. H. III. Prior to such a discussion it is useful to make some general points concerning the relationship between the π-ligand functions and the spherical harmonic functions discussed above. In a co-ordination polyhedron the N ligands, each individually having cylindrical symmetry, have a total of 2N orbitals which have π-symmetry with respect to the metal-ligand bond axes. Clearly such orbitals cannot be described in terms of simple spherical harmonic functions, because π-orbitals are noded at the ligand positions. Stone[18] in a very important and elegant paper has demonstrated that such functions can, however, be described in terms of vector surface harmonic functions. The mathematical aspects of this treatment have been given by Stone; we summarise the more qualitative aspects here. The important points relevant to an understanding of the ligand π-functions are as follows:

(1) The spherical harmonics provide a set of scalar functions which are solutions to the particle on the sphere problem. In addition, however, there are solutions which satisfy the Schrodinger equation and are based on vectors. Two such functions may be constructed from each of the scalar spherical harmonics with the exception of Y_{00}. One of these functions is said to be of even parity, i.e. it shows the same behaviour under inversion as the parent scalar harmonic function $Y_{l,m}$ and is denoted $V_{l,m}$. The other function has odd parity, i.e. it behaves in the opposite manner under the inversion operation and is denoted $\overline{V}_{l,m}$. The same l and m quantum numbers are used to describe these functions.

(2) The magnitude and direction of the even parity vector harmonic function, $V_{l,m}$, at the co-ordinates (θ, ϕ) represent the gradient of the parent scalar harmonic $Y_{l,m}$ at that point. The corresponding odd parity vector harmonic, $\overline{V}_{l,m}$, has the same magnitude as $V_{l,m}$ but the vectors are rotated by 90° in an anticlockwise fashion when viewed down the

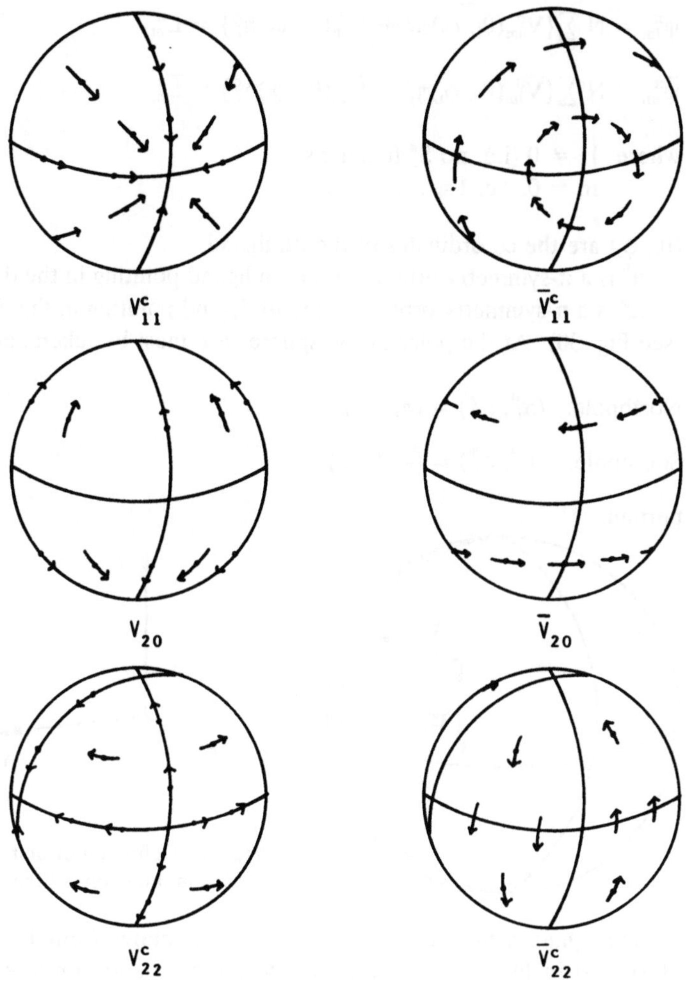

Fig. 29. Relationship between even and odd vector spherical harmonics

M–L bond axis. The realtionship between the even and odd vector surface harmonics is illustrated in Fig. 29.

(3) At any point (θ, ϕ) the vector functions may be written in terms of two orthogonal vector components – one pointing in the direction of increasing θ and the other in the direction of increasing ϕ.

i.e. $V_{l,m} = (V^{\theta}_{l,m}, V^{\phi}_{l,m})$

$\bar{V}_{l,m} = (\bar{V}^{\theta}_{l,m}, \bar{V}^{\phi}_{l,m})$

(4) The ligand π-functions may be expressed by the following expansions which are analogous to those used for the L^{σ} functions:

$$\psi^{\pi}_{lm} = N \sum\{V^{\theta}_{lm}(\theta_i, \phi_i)\pi^{\theta}_i + V^{\phi}_{lm}(\theta_i, \phi_i)\pi^{\phi}_i\} = L^{\pi}_m$$

$$\overline{\psi}^{\pi}_{lm} = N' \sum\{\overline{V}^{\theta}_{lm}(\theta_i, \phi_i)\pi^{\theta}_i + \overline{V}^{\phi}_{lm}(\theta_i, \phi_i)\pi^{\phi}_i\} = \overline{L}^{\pi}_m \qquad \text{(VIII)}$$

where $l \neq 0$ (i.e. no S^{π} functions)
$m = 0, 1c, 1s, \ldots\ldots. lc, ls$

(θ_i, ϕ_i) are the co-ordinates of the ith ligand.

π^{θ}_i is a π-symmetry orbital on the ith ligand pointing in the direction of increasing θ.

π^{ϕ}_i is a π-symmetry orbital on the ith ligand pointing in the direction of increasing ϕ (see Fig. 30). At the poles of the sphere care must be taken and the relationships:

$$\text{Northpole:} \quad (\pi^{\theta}_i, \pi^{\phi}_i) = (\pi^{x}_i, \pi^{y}_i)$$

$$\text{Southpole:} \quad (\pi^{\theta}_i, \pi^{\phi}_i) = (-\pi^{x}_i, \pi^{y}_i) \qquad \text{(IX)}$$

pertain.

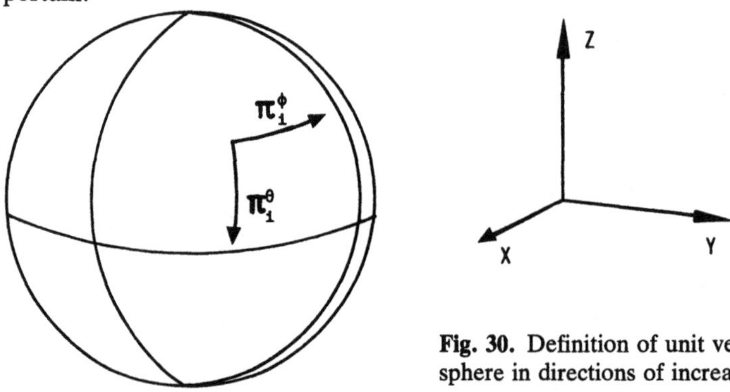

Fig. 30. Definition of unit vectors on the surface of a sphere in directions of increasing θ and ϕ

The ligand π-functions generated from the above formulae for the octahedron are illustrated in Fig. 31. Only the first N non zero and non-repeating ligand L^{π} and $\overline{L}^{\pi}$ functions are used.

(5) The ligand functions denoted by $\overline{L}^{\pi}_m$ and generated from the odd parity $\overline{V}_{l,m}$ vector functions may also be generated from even parity $V_{l,m}$ vector functions, with in general higher l quantum numbers. This observation has been exploited by Quinn[17] in his generator orbital approach.

Consider a linear $D_{\infty h}$ MX_2 system. The following ligand π-functions are generated from the vector surface harmonics.

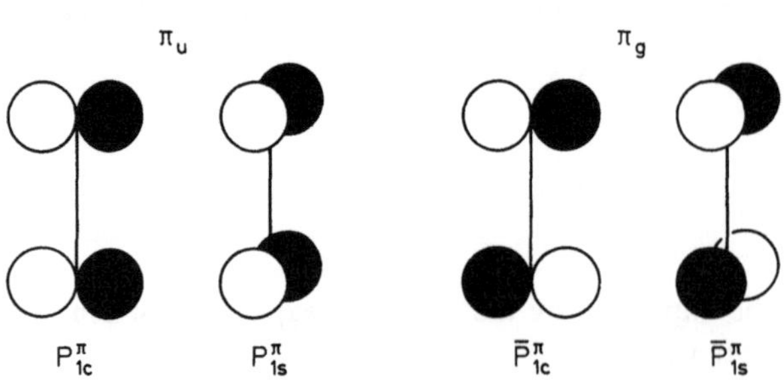

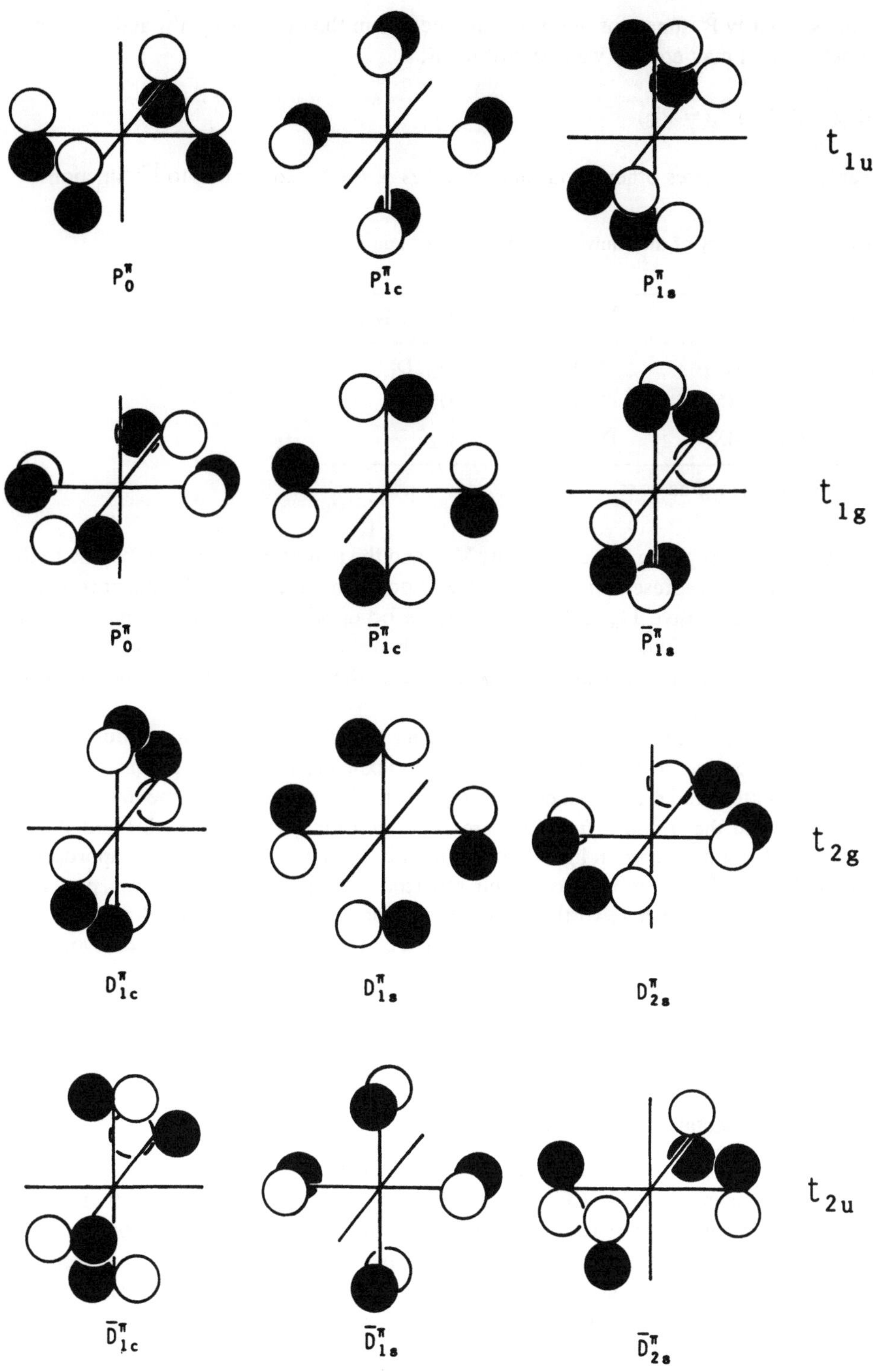

Fig. 31. π-functions for an octahedral complex

The odd parity $\bar{P}^{\pi}$ functions are also generated from the even parity V^{c}_{21} and V^{s}_{21} vector functions and may equally well be written as:

$D^{\pi}_{1c}(=\bar{P}^{\pi}_{1s})\quad D^{\pi}_{1s}(=\bar{P}^{\pi}_{1c})$

Table 13 summarises other important instances of the L^{π} equivalent to $\bar{L}^{\pi}$ functions.

Table 13. Summary of the equivalences between $\bar{L}^{\pi}$ and L^{π} functions

Structure		$\bar{L}^{\pi}$	L^{π} Equivalent
Linear	MX_2	$\bar{P}^{\pi}_{1c}, \bar{P}^{\pi}_{1s}$	$D^{\pi}_{1s}, D^{\pi}_{1c}$
Planar	MX_n	$\bar{P}^{\pi}_{1c}, \bar{P}^{\pi}_{1s}$	$D^{\pi}_{1s}, D^{\pi}_{1c}$
Tetrahedron	MX_4	$\bar{D}^{\pi}_{0}$	D^{π}_{2c} [a]

[a] One component of an e set

(6) For any point group, the even parity $V_{l,m}$ and the parent $Y_{l,m}$ functions belong to the same irreducible representation (and hence so do the related L^{σ}_{m} and L^{π}_{m} linear combinations). Thus extensive L^{σ}_{m} and L^{π}_{m} mixing can occur when the quantum numbers are matched in this fashion. This also means that any destabilisation of filled metal d orbitals by secondary P^{σ}–d mixings (for example the t_2 set in the tetrahedron) are mitigated by the P^{π} orbitals of π-acceptor ligands.

As a result of the vector surface harmonic functions having zero modulus at co-ordinates where the scalar functions have their maxima (i.e. when the partial derivatives with respect to θ and ϕ are both equal to zero) some interesting complementary relationships between the σ and π-functions are generated for ML_N. The P^{σ}, D^{σ} and P^{π}, D^{π} manifolds for some common ligand polyhedra are given in Table 14. Where appropriate the odd parity $\bar{L}^{\pi}$ functions have been replaced by higher order and even parity L^{π} functions. Particularly noteworthy is the observation that for *closo*-polyhedra with $N > 5$ all the vacancies in the D^{σ} manifold are matched by D^{π} functions with identical symmetry characteristics.

II. π-Acceptor Ligands

a. Transition Metals

The ability of transition metals to co-ordinate molecules with low dipole moments and high ionisation energies, e.g. CO, NO, N_2 and PF_3, represents one of their more interesting and characteristic properties[79]. Indeed the superior qualities of transition metals as homogeneous and hetereogeneous catalysts is dependent in large measure on this chemical property. Although such ligands are not effective σ-donors they have low lying π-acceptor orbitals which are able to stabilise filled orbitals on the central metal atom with matching symmetry characteristics. For cylindrically symmetric π-acceptor ligands 2N π-functions are generated and in general they are capable of matching filled non-bonding metal orbitals both in their symmetries and nodal characteristics. The surfeit of

Table 14. P^{σ}, D^{σ} and P^{π}, P^{π} functions for common co-ordination polyhedra. L^{σ} linear combinations are represented by the hatched lines and L^{π} by the filled lines

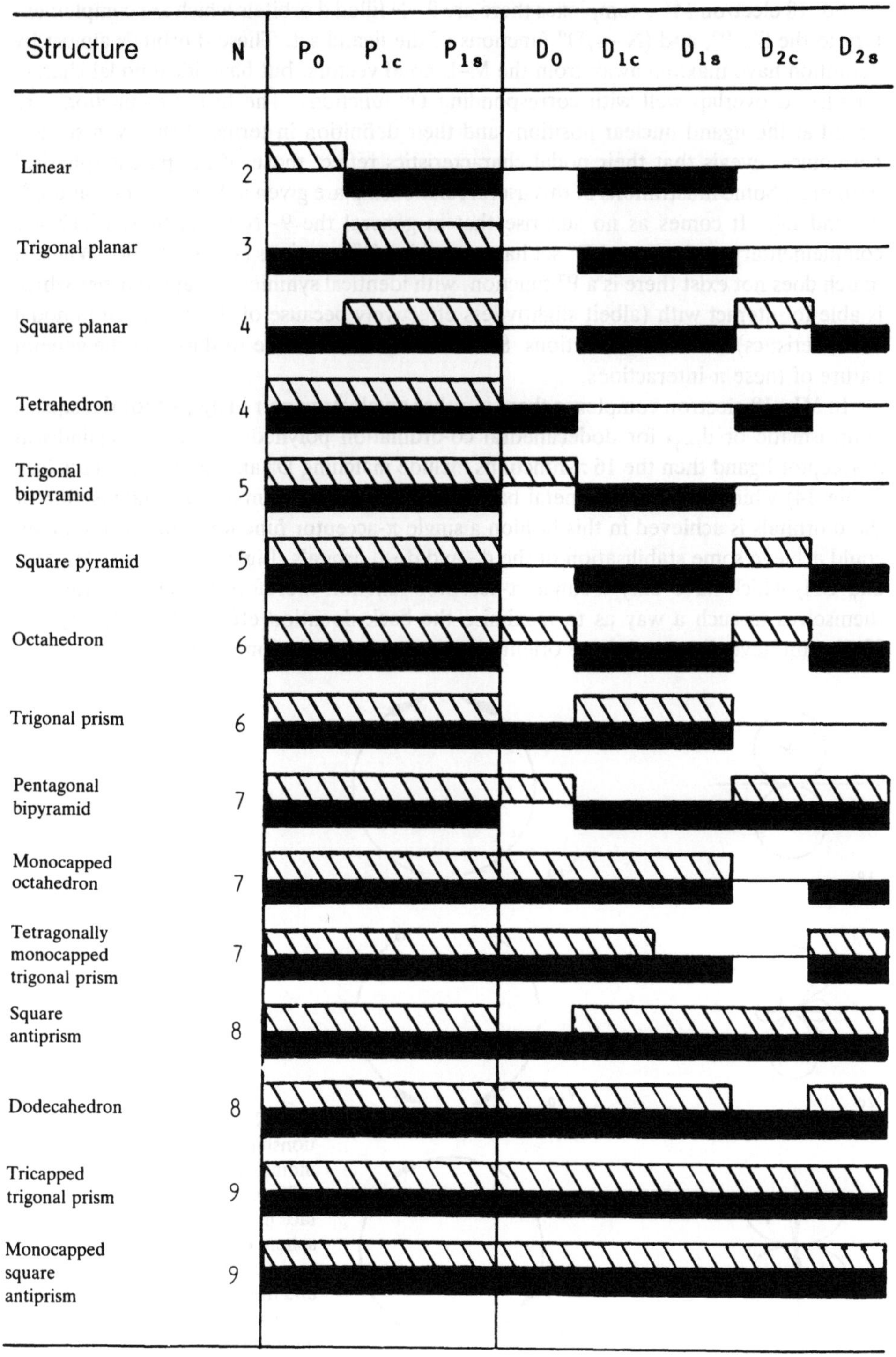

π-functions generated is emphasised by Table 14 which summarises the P^π and D^π functions for some common co-ordination polyhedra.

For 18 electron ML_N complexes there are $9-N$ filled d orbitals which are complementary to the S^σ, P^σ, and $(N-4)D^\sigma$ functions of the ligand set. These d orbitals almost by definition have maxima away from the M–L bond vectors, but have ideal nodal characteristics to overlap well with corresponding D^π functions. The ligand π-functions are noded at the ligand nuclear positions and their definition in terms of the even surface harmonics reveals that their nodal characteristics reflect those of the parent spherical harmonic. Some illustrations of this useful relationship are given in Fig. 32 for d_{z^2} and D_0^π, d_{xz} and D_{2c}^π. It comes as no surprise that in general the $9-N$ d functions which are complementary to the ligand D^σ set have matching D^π functions (see Table 14). Where a match does not exist there is a P^π function, with identical symmetry characteristics which is able to interact with (albeit slightly less effectively because of the mismatch in nodal characteristics) with the d functions. Some examples will serve to illustrate the general nature of these π-interactions.

In ML_8 18 electron complexes there is either an electron pair in d_{z^2} (d_0) for the square antiprismatic or $d_{x^2-y^2}$ for dodecahedral co-ordination polyhedra. If L is a cylindrical π-acceptor ligand then the 16 π-functions include matching D_0^π and D_{2c}^π components (see Table 14) which stabilise the metal based orbitals. Although maximum stabilisation of the d orbitals is achieved in this fashion a single π-acceptor function suitably orientated could achieve some stabilisation of the d_{z^2} and $d_{x^2-y^2}$ orbitals. Ligands such as C_2H_4, SO_2 and CH_2 which have only a single π-acceptor function therefore locate and orientate themselves in such a way as to maximise the back donation effect. Burdett, Fay and Hoffmann have discussed these orientational preferences in some detail[35].

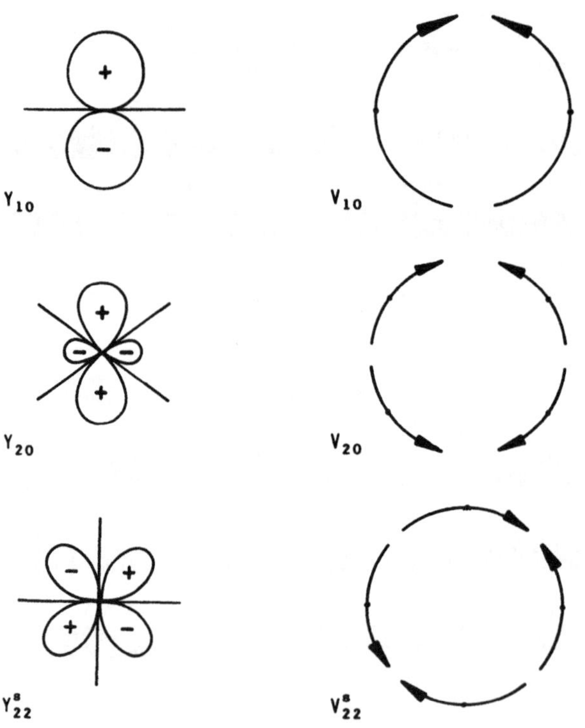

Fig. 32. Illustrations of the relationships between spherical harmonics on the central atom (*left column*) and matching vector surface harmonics on the ligand sphere (*right column*). The *arrows* are representative of p functions and their direction reflects that of the gradient of the parent spherical harmonic

The complementary filled d orbitals in a pentagonal bipyramidal ML_7 complex are d_{xz} and d_{yz} (d_{1c}, d_{1s}) and are therefore stabilised by D^{π}_{1c} and D^{π}_{1s} π-acceptor linear combinations. In the limit a single cylindrically symmetrical π-acceptor ligand located in an axial position (but not equatorial) would match this orbital pair (see Fig. 33). A pair of single faced π-acceptor ligands in axial positions would adopt a staggered disposition to maximise their interactions with d_{xz} and d_{yz}.

In the case of the octahedron there is a perfect match between the non-bonding d_{xz}, d_{yz} and d_{xy} orbitals and the complementary D^{π}_{1s}, D^{π}_{1c} and D^{π}_{2s} orbitals of the π-acceptors. Indeed the wide occurence of the octahedral stereochemistry can be attributed to the fact that it represents simultaneously the best solutions to the sphere packing and covering problems and provides the most effective situation for π-bonding effects. Indeed sometimes the octahedral geometry is retained although it requires antibonding interactions between S^{σ} and a filled s orbital on the central atom, e.g. $[TeCl_6]^{2-}$.

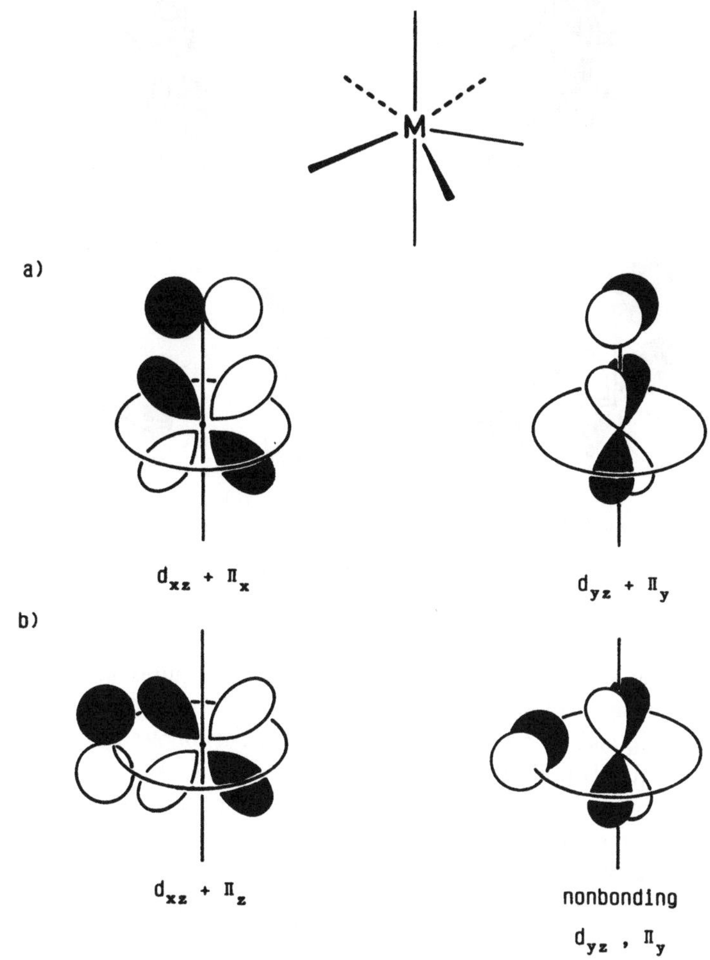

Fig. 33 a, b. π-bonding effects in a pentagonal bipyramidal complex. In (**a**) an acceptor ligand with cylindrical symmetry is located in an axial position and in (**b**) an equatorial position. In the latter the symmetries of d_{yz} and π_y do not match

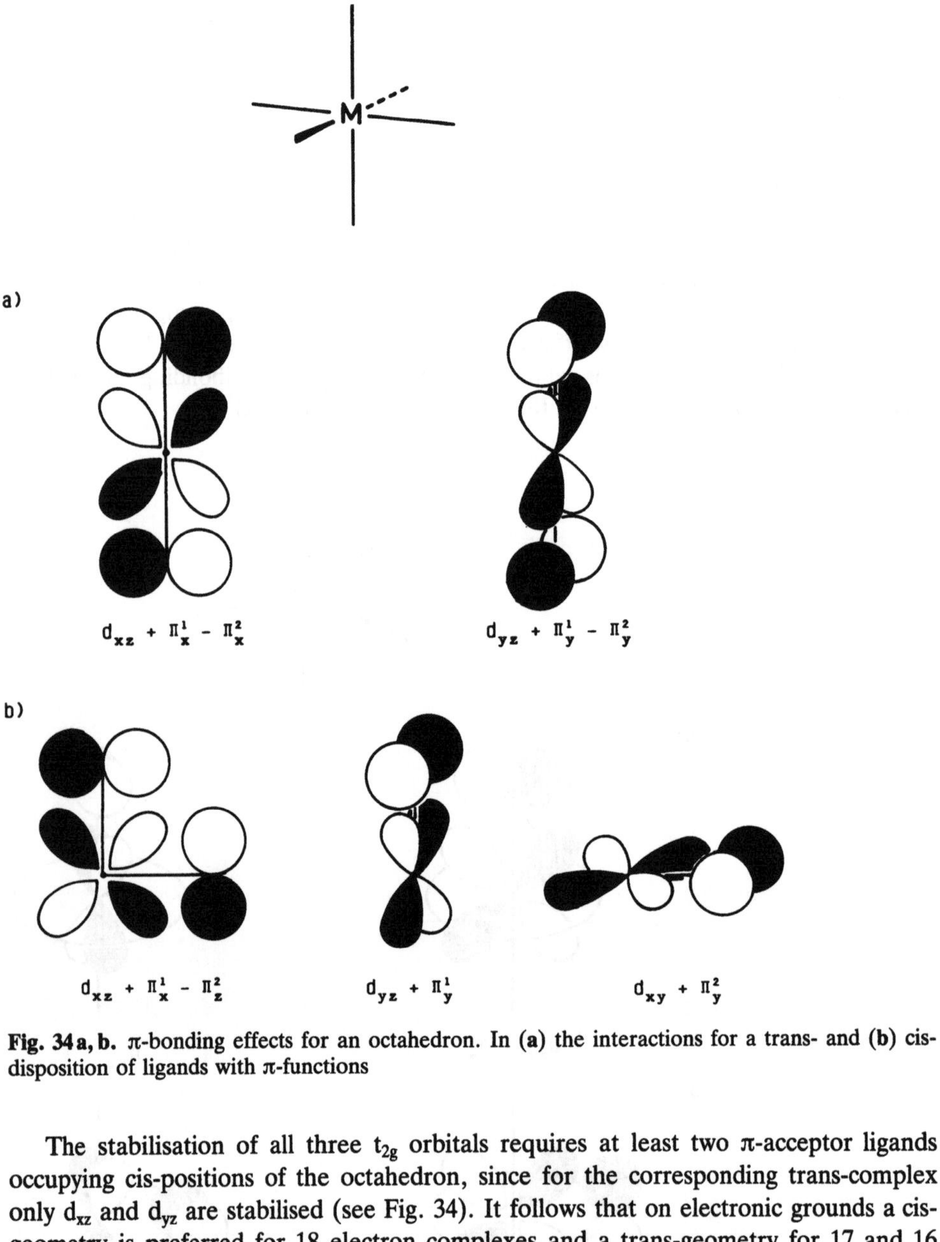

Fig. 34 a, b. π-bonding effects for an octahedron. In (**a**) the interactions for a trans- and (**b**) cis-disposition of ligands with π-functions

The stabilisation of all three t_{2g} orbitals requires at least two π-acceptor ligands occupying cis-positions of the octahedron, since for the corresponding trans-complex only d_{xz} and d_{yz} are stabilised (see Fig. 34). It follows that on electronic grounds a cis-geometry is preferred for 18 electron complexes and a trans-geometry for 17 and 16 electron complexes. This serves as the basis for the general conclusion that in 18 electron complexes π-acceptor ligands adopt cis-dispositions if these are permitted by the steric requirements of the other ligands[80].

In a square-pyramidal complex, ML_5, an examination of Table 14 suggests that although d_{xz}, d_{yz} and d_{xy} are capable of stabilisation by π-acceptor functions d_{z^2} is not. This is only true if all the L–M–L angles = 90°. If the equatorial ligands are bent back in such a way that C_{4v} symmetry is maintained then overlap between d_{z^2} and P_0^{π} is permitted (see below).

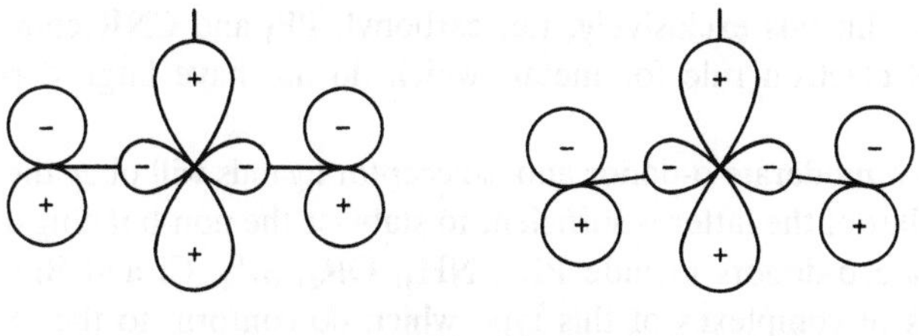

In ML_5 transition metal complexes this angle is usually found to lie between 93° and 105°.

In the corresponding trigonal bipyramidal complex D^{π}_{1s} and D^{π}_{1c} ligand π-acceptor functions stabilise the non-bonding d_{xz} and d_{yz} orbitals, but the weakly anti-bonding d_{xy} and $d_{x^2-y^2}$ orbitals do not have matching D^{π} functions. The P^{π}_{1s} and P^{π}_{1c} functions have matching symmetry characteristics and therefore d–p mixing permits a diminution of the anti-bonding character of d_{xy} and $d_{x^2-y^2}$ and encourages some back donation (see Fig. 35).

For tetrahedral ML_4 the ligand π-functions generate only D^{π}_0 and D^{π}_{2c} functions to match d_{z^2} and $d_{x^2-y^2}$ (e). The d_{xz}, d_{yz} and d_{xy} (t_2) have the same symmetry as P^{π}_0, P^{π}_{1s} and P^{π}_{1c} and are therefore stabilised by d–P^{π} mixing in much the same manner as that described above for the trigonal bipyramid.

From the analysis developed above and that given in Sect. D the following guidelines emerge for eighteen electron compounds:

(a) Strong σ-donor ligands which are able to overlap effectively with the metal d, s and p valence orbitals form complexes which adhere to the 18 electron rule. In view of the large differences in the radial distribution functions of these orbitals small ligands such as H which are able to come sufficiently close to the metal to overlap with the d orbitals are particularly effective in this respect. Examples, of hydrido- and hydrido-phosphine complexes which conform to the inert gas rule are summarised in Tables 5–9.

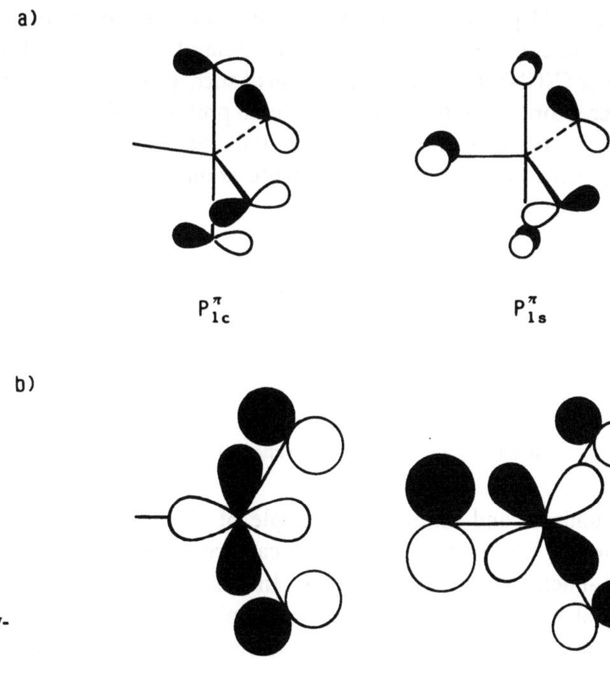

Fig. 35 a, b. Illustration of p mixing effects in a trigonal bipyramidal complex

(b) Complexes with π-acceptor ligands exclusively, i.e. carbonyl, PF_3 and CNR complexes will conform to the 18 electron rule for metals which do not have large d–p promotions energies.
(c) Mixed metal complexes with moderate σ-donor and π-acceptor ligands will obey the 18 electron rule when the number of the latter is sufficient to stabilise the non-bonding d based metal orbitals. (Moderate σ-donors include PR_3, NH_3, OR_2, SR_2, Cl and Br). There are numerous examples of complexes of this type which do conform to the 18 electron rule.
(d) The formation of metal-metal bonds in odd electron low valent carbonyl and related complexes is also important in influencing the viability of the 18 electron rule. The whole question of metal-metal bonding in this context will be described in a subsequent review, suffice to say that as the metal-metal bonding gets stronger, i.e. down the periodic table, fewer exceptions to the 18 electron rule are found.

b. Main Group Analogues

Although it is not common for π-acid ligands to co-ordinate to main group atoms it is instructive to consider some illustrative cases in order to explore the stereochemical ramifications. For example C_3O_2[81] can be formally viewed as a carbonyl complex of carbon, OC–C–CO, and therefore is directly analogous to the carbonyl complexes described above. The linear arranement of carbonyl ligands generates S^σ and P_0^π lone pair linear combinations, which donate electron density into empty s and p_z orbitals on carbon. Four electrons occupy the carbon $2p_x$ and $2p_y$ orbitals which are stabilised substantially by overlap with the empty π* orbitals of the carbonyls in much the same way as that described for $Mo(CO)_6$. However, for main group molecules the ligand polyhedra which are suitable for back-donation effects of this type are limited to linear and trigonal planar, since only these geometries are consistent with the presence of filled orbitals with exclusively p orbital character. In the trigonalplanar compounds there is a p_z orbital perpendicular to the ligand plane capable of overlap with the π* orbitals of the ligands. Planar $C(CN)_3^-$ provides an example of such a compound. If the ligands have only a single π-acceptor orbital then conformational preferences which maximise the overlap between this orbital and the filled p orbitals on the central atom are observed. Allene, $H_2C{=}C{=}CH_2$ and $(CH_2)_3X^{2+}$ provide examples of these conformational effects and their structures are illustrated below[82].

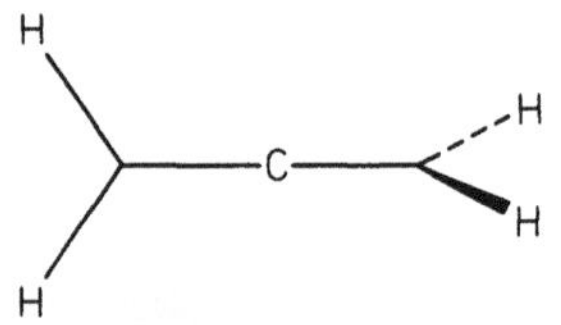

allene

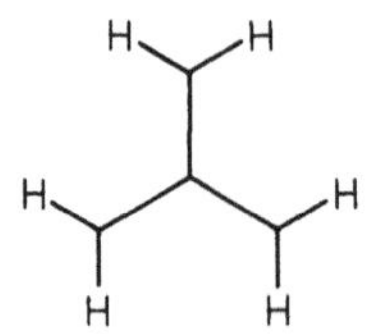

trimethylenemethane dication

An unusual and interesting example of a π-acid ligand in the context of main group chemistry is provided by Li, which can accept electron density into its empty 2p orbitals. OLi_2-linear provides an example. In this compound the octet rule is obeyed about the central atom. Von Schleyer has made a detailed and elegant investigation of such compounds using ab initio theoretical techniques[83–85]. Interestingly, compounds of the type

CLi_6 at first examination appear to be exceptions to the inert gas rule, but he has shown that these „hyper-metalated compounds" have only 8 electrons occupying S^σ and P^σ type molecular orbitals around the central atom. The remaining electrons occupy orbitals which are localised predominantly on the lithium atoms, are outpointing and weakly metal-metal bonding.

This raises the very interesting question of why OLi_2 is linear whereas OH_2 is angular? In the former case the additional electron pairs in O $2p_x$ and $2p_y$ are stabilised by back donation to the lithium orbitals. In OH_2 these effects are absent and the electrons are localised completely on the oxygen atom. Since the 2s orbitals of oxygen have greater orbital ionization energies a distortion which introduces more s character into these orbitals is energetically favourable[10]. There is not a corresponding change in geometry in going from transition metal π-acceptor complexes to isoelectronic hydrido-complexes, e.g. $Mo(CO)_6$ to RuH_6^{2-} because the electrons are in both instances localised predominantly in metal d orbitals, i.e. the orbitals with the highest orbital ionisation energies. This suggests an interesting compensation effect between the ligand and central atom spheres. For main group atoms more s electron density is localised on the central atom and more p character on the ligand sphere. For transition metal atoms more d orbital character is concentrated on the central atom and compensated for by more S and P character on the ligand sphere.

As the π-acceptor character of the ligands in main group compounds decreases then it becomes finely balanced whether the linear or angular geometries are preferred. The soft potential energy surfaces associated with $(Ph_3P)_2N^+$ and the angular geometry of $(Ph_3P)_2C$[86] can be interpreted in terms of the poor acceptor character of PPh_3.

III. π-Donor Complexes

π-donor complexes based on O, N and F ligands are a feature not only of the earlier transition metals but also of the main group elements. In these complexes the π-donor functions are localised predominantly on the ligands and require empty orbitals on the central atom to stabilise them. The symmetry and nodal characteristics of the problem are identical with those discussed above, but the chemical consequences are quite different. Since these ligands have high electronegativities the filled π-donor orbitals are lower lying than the non-bonding d-orbitals and a stabilising interaction results only if the metal orbitals are empty. The contrasting situations for π-donor and π-acceptor ligands are shown in Fig. 36. It follows that as long as the filled ligand orbitals match the non-bonding d orbitals in symmetry and the overlap between the orbitals is large then the d-orbitals are destabilised by the interaction, i.e. d^0 configurations are preferred if the conventional oxidation state formalism is followed. In this context effective π-donor ligands include O^{2-}, N^{3-} and NR^{2-}. Ligands which are less effective π-donors, e.g. F^-, Cl^-, etc., do not destabilise the non-bonding d orbitals to such a great extent and population of the d orbitals in such situations is a common occurence.

Although in the previous sections the importance of achieving a stable and spherical electron distribution about the central atom has been stressed as a major contributing force behind the 18 electron rule formalism, there are in fact interlocking spherical regions of electron density in any complex each of which is trying to attain an inert gas situation. For example, in a complex such as MnO_4^- each of the oxygen atoms attains an

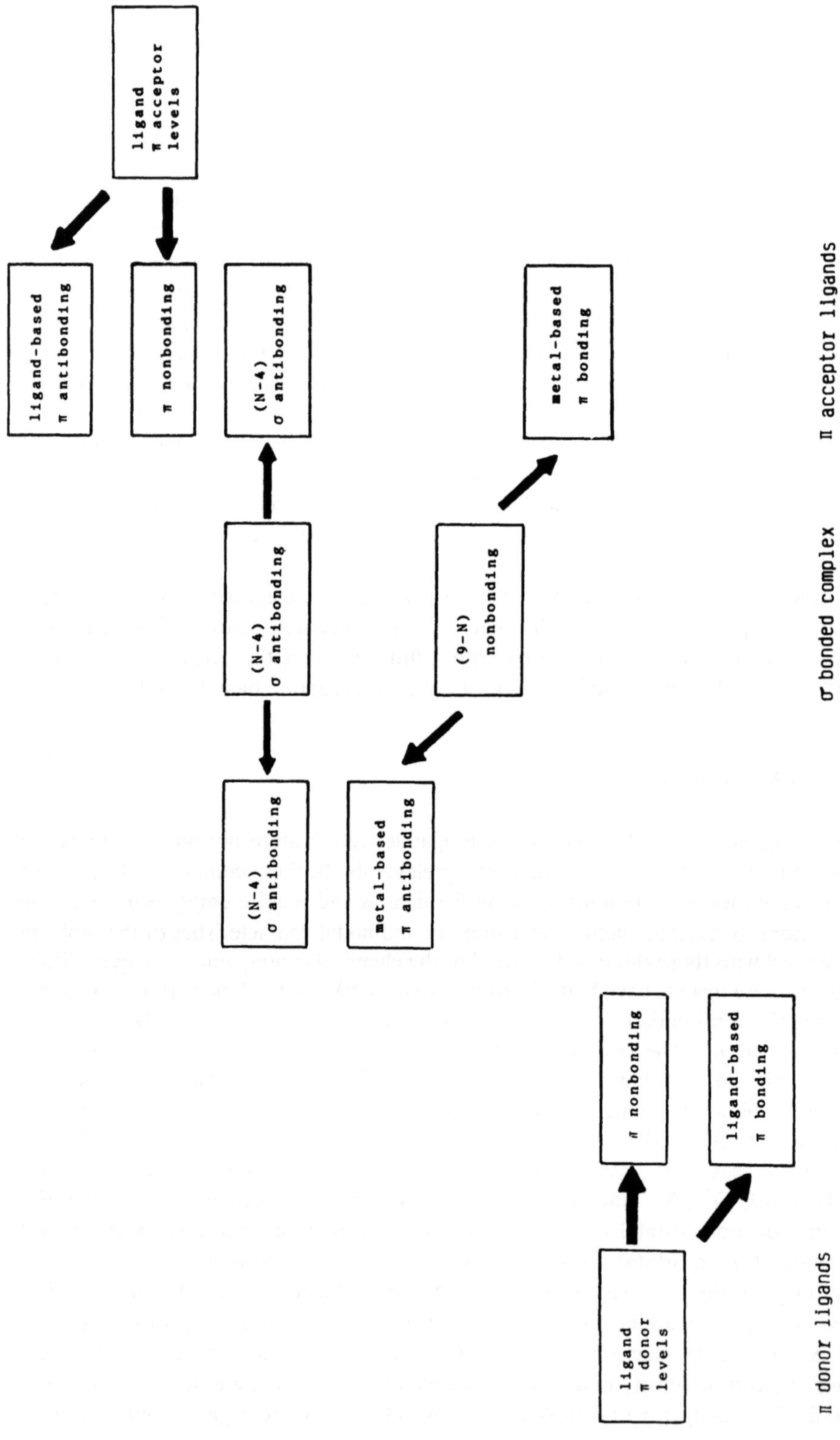

Fig. 36. Comparison of π-acceptor and donor effects in transition metal complexes

inert gas configuration by means of overlap with the manganese 3d, 4s and 4p orbitals. Since oxygen is more electronegative than Mn the resultant molecular orbitals are localised more extensively on O and are stabilised by empty orbitals on the central atom, i.e. the situation is the converse of that described in the context of π-acceptor ligands, where filled orbitals on the central atom are stabilised by empty orbitals on the ligands. Consequently, in π-donor complexes the driving force is the acievement of the inert gas configuration for the peripheral atoms[87,88] and a spherical electron distribution for the central metal atom. This leads to the following simple rules for main group and transition metal π-donor complexes, ML_N.

(a) Main group compounds have the following characteristic geometries as a function of electron count: 8N: *closo*-, 8N + 2: *nido*-, 8N + 4: *arachno*-, 8N + 6: *hypho*-.

(b) Transition metal complexes with 8N valence electrons have *closo*-geometries.

The total valence electron count is of course made over all atoms and for example in CO_2 it is 4(C) + 2 × 6(O) = 16. Examples, of main group compounds which conform to this generalisation are given in Table 1a. The *closo*-, *nido*-, *arachno*-, designation follows from the need to achieve spherical electron distributions about the central atom and corresponds to the analysis for MH_N, $MH_{N-1}E$, $MH_{N-2}E_2$, ... etc given in Sect. D. A similar series is not observed for transition metals because additional electron pairs do not result in the generation of *nido*-, *arachno*-, etc. structures, but rather promote ligand dissociation.

The view that the peripheral atoms are stabilised by virtual orbitals on the central atom rationalisises the observation that electronegative atoms generally adopt locations on the periphery of the molecule in preference to the central position[15]. Interestingly, it is also possible to get compounds with 18N valence electrons where a central main group atom is co-ordinated to transition metal "ligand" fragments and some examples are given in Table 15[88-91].

Table 15. Some examples of main group compounds with σ-donor transition metal ligands having 18N valence electrons

Molecular formula	Coordination about central nonmental atom	Structure	n_e
$[Ru_2OCl_{10}]^{4-}$	$O(RuCl_5)_2$	linear	36
$[Ru_2NCl_8(H_2O)_2]^{3-}$	$N\{RuCl_4(H_2O)\}_2$	linear	36
$[Ir_3N(SO_4)_6(H_2O)_3]^{4-}$	$N\{Ir(SO_4)_2(H_2O)\}_3$	trigonal planar	54

I. Summary

The complementary relationships which have been developed for main group and transition metal covalent compounds can be summarised in the economical fashion illustrated in Fig. 37. The central atom sphere is represented by the inner circle and the ligand donor orbitals by the outer segments. The localisations of the molecular orbitals are indicated in an approximate fashion by the shadings – a shaded area representing molecular orbitals localised predominantly on the metal or ligands and the unshaded areas empty orbitals localised in the corresponding fashion. The onset of donation from filled orbitals to empty orbitals will of course lead to a more spherical electron distribution than that indicated schematically in the Figure. Elementary molecular orbital considerations suggest that filled orbitals on the ligands must be matched by empty orbitals on the central atom and vice versa. The Figure summarises this aspect and also indicates the stereochemical preferences for the important classes of main group and transition metal covalent compounds.

1. Main Group Hydrides, MH_N – Figure 37a

The ligand linear combinations together with a matching set of equivalent orbitals on the central atom generate four S^σ and P^σ functions and are therefore characterised by a total of 8 valence electrons, e.g. CH_4, NH_3, OH_2 and HF. The packing and covering solution of the sphere require that the electron density be localised in a tetrahedral fashion. Planar and linear stereochemistries are not favoured for NH_3 and OH_2 because electron pairs are localised exclusively in 2p orbitals which have lower valence orbital ionisation energies than the 2s orbitals. The distortions away from the idealised tetrahedral bond angles in NH_3 and OH_2 result from a redistribution of electron density which places more s character on the equivalent orbitals localised on the central atom and more P^σ orbital character on the spherical harmonic linear combinations of hydrogen 1s orbitals.

2. Transition Metal Hydrides, MH_N – Figure 37b

The hydrogen linear combinations generate S^σ, P^σ and $(N-4)D^\sigma$ functions if they utilise polyhedra which are solutions to the packing and covering problems. The $(9-N)$ filled d orbitals form a complementary set. When taken together the ligand linear combinations and the metal orbitals form a set of orbitals whose nodal and radial characteristics approximate to those of an inert gas. Therefore such molecules are characterised by 18 valence electrons, e.g. $[ReH_9]^{2-}$, $[RuH_6]^{4-}$ and $[NiH_4]^{4-}$.[92].

3. Transition Metal π-Acceptor Complexes, MA_N – Figure 37d

The lone pairs of the π-acceptor ligands generate S^σ, P^σ and $(N-4)D^\sigma$ functions as in 2. above, but the $(9-N)$ complementary d orbitals are stabilised by back donation from these orbitals into P^π/D^π functions derived from virtual orbitals of the π-acceptor ligands. These complexes adhere strongly to the eighteen electron rule as a consequence of the complete matching of ligand and metal orbitals illustrated in the Figure. Examples include $Cr(CO)_6$, $Fe(CO)_5$ and $Ni(CO)_4$. Other ligands which have good π-acceptor properties are PF_3, CNR, NO and CS.

4. Transition Metal π-Acceptor Complexes With Fewer Than 18 Valence Electrons MA_NE – Figure 37c

These complexes adopt *nido-*, *arachno-* and *hypho-* geometries derived from the polyhedra described in 3, if they have 16, 14 and 12 electrons respectively. The vacancies in the co-ordination polyhedra are characterised by outpointing empty equivalent orbitals which point towards the missing vertices and have the maximum amount of p orbital character. Examples include, $[Ni(CN)_4]^{2-}$ – square-planar (16 electron), and $[Rh(PPh_3)_3]^+$ – T-shaped (14 electron) which are both derived from the square pyramid. When the ligands are less effective π-acceptors then alternative structures based on the octahedron, pentagonal bipyramid and dodecahedron with electron pair holes in non-

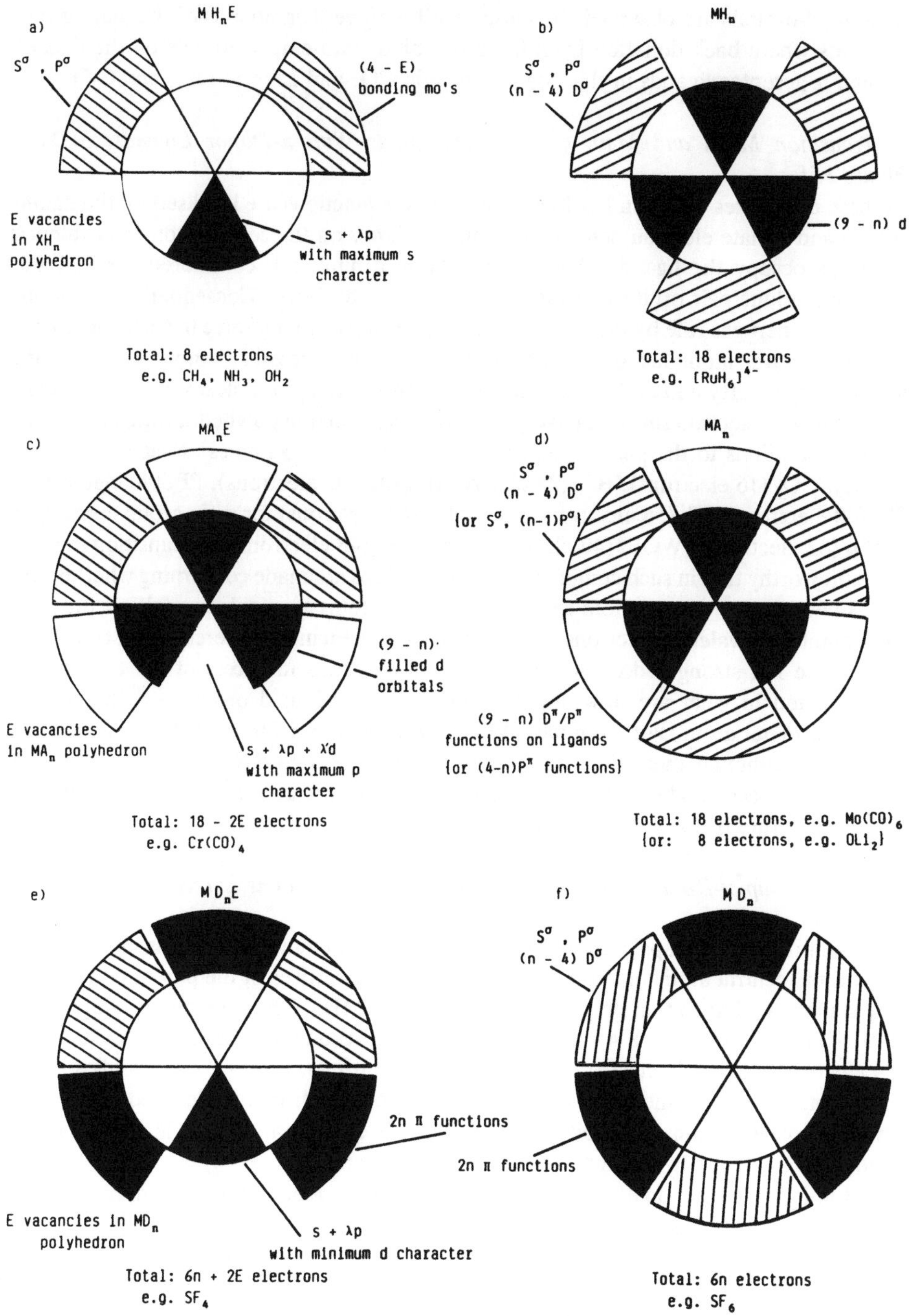

Fig. 37 a–f. Summary of complementary relationships in main group and transition metal compounds – (**a**) main group hydrides, (**b**) transition metal hydrides, (**c**) coordinatively unsaturated transition metal compounds, (**d**) 18 electron (transition metal) and 8 electron (main group) compounds with π-acceptor ligands, (**e**) main group π-donor complexes and (**f**) main group and transition metal π-donor complexes with deltahedral geometries bonding d-orbitals are observed. It is also possible to get compounds of the main group elements where back donation from filled p orbitals into empty orbitals on the ligands occurs. π-acceptor ligands in this context include CO, and Li, e.g. OLi_2 and C_3O_2

5. Transition Metal and Main Group Compounds With π-Donor Ligands, MD_N – Figure 37f

In these complexes the ligand σ-donor and π-donor functions are localised on the ligand sphere and donate electron density into empty orbitals on the central atom. 6 N valence electrons occupy the ligand sphere if only these orbitals are considered – 8 N if the outpointing lone pair orbitals on the D ligands are considered. Consequently the adoption of the inert gas rule by the ligand atoms is a primary driving force in such complexes and the central atom facilitates this by providing the necessary empty orbitals with matching symmetry and nodal characteristics. Consequently complexes of this type conform to an 8 N electron rule and have geometries based on the polyhedra which represent the best solutions to the packing and covering problems. Examples of such complexes include, CO_2 (16 electrons), BF_3 (24 electrons), SO_4^{2-} (32 electrons), PF_5 (40 electrons), SF_6 (48 electrons), IF_7 (56 electrons) and XeF_8^{2-} (64 electrons) for main group atoms and MnO_4^- (32 electrons), WOF_4 (40 electrons) and WF_6 (48 electrons) for transition metals. It is noteworthy that in such complexes no suppositions are made concerning whether the bonds between the central atom and the peripheral atoms are single or multiple only the total number of valence electrons is important in influencing the stereochemistry. If the ligands are not strong π-donors then additional electrons in excess of the 8 N can be accomodated in what are now approximately non-bonding d orbitals on the central transition metal atom. For example, WF_6, ReF_6, OsF_6 and PtF_6[93] all have octahedral geometries although their electron counts vary from 8 N to 8 N + 4. The additional electrons are accomodated in the t_{2g} orbitals which are approximately non-bonding because fluorine is not a strong π-donor.

6. Main Group π-Donor Complexes Where the Electron Count Exceeds 8 N, MD_NE – Figure 37e.

The additional electron pairs in such compounds occupy equivalent orbitals mainly localised on the central atom and pointing towards missing vertices of the parent polyhedron in a manner similar to that described in 1. above for MH_N. The compounds adopt *nido*-(8 N + 2), *arachno*- (8 N + 4), and *hypho*- (8 N + 6) geometries. Examples include SF_4 (34 electrons), ClF_3 (28 electrons) and XeF_2 (22 electrons). If E = 2 the ligands adopt a planar and if E = 3 a linear arrangement. The distortions from the idealised geometries reflect a concentration of s orbital character in the equivalent orbitals localised on the central atom and D^σ in the ligand linear combinations.

Although a distinction has been made above between π-acceptor and π-donor ligands which follows the usual conventions of oxidation state formalisms, these are just formalisms and represent chemically sensible ways of partitioning electron density. There are situations where a ligand in a complex may be formulated either as a π-donor or π-acceptor. For example, a carbyne complex $W(CO)_5CR^+$[94] may be described either a complex of CR^{3-} leading to a d^2 complex (W(IV)) where it is functioning as a π-donor donating electron density into empty d_{xz} and d_{yz} orbitals, or as a complex of CR^+ leading to a d^6 (W(O)) complex, where it is functioning as a π-acceptor accepting electron density from filled d_{xz} and d_{yz} orbitals. Of course neither is strictly true and the real situation is somewhere between the extremes. Since there are alternative electron counting rules for complexes where π-acceptors or π-donors predominate it makes sense to choose the formalism consistent with the larger number of ligands, i.e. in the above example treat CR as a π-acceptor ligand.

The salient point to recognise is that the electron counting rules described above represent a way of enumerating the number of antibonding orbitals unavailable for electron occupation[95,96]. It matters not whether these molecular orbitals are localised more on the metal or the ligand for purposes of electron counting. Figure 37 provides an easy method of establishing the number of antibonding orbitals of this type. For example, in $W(CO)_6$ each carbonyl ligand has two antibonding π^* and an antibonding σ^* orbital. From (d) it is evident that these are retained in the complex, i.e. there are 18 such orbitals. In addition six new antibonding orbitals result from complex formation and in particular from the σ-bonds. Therefore, from the 57 valence orbitals in the molecule 24 become antibonding and a total of 66 valence electrons can be accomodated in bonding molecular orbitals. In a complex such a WO_4F_2 it is evident from Fig. 37 (f) that there are a total of 9 antibonding molecular orbitals and therefore the 33 valence orbitals give rise to a total of 24 bonding molecular orbitals which can accommodate 48 electrons. This process of identifying the number of antibonding orbitals might prove useful in describing the electron counts where the situation is ambiguous. For example, in $W(PhC{\equiv}CPh)_3(CO)$ the total number of electrons exceeds that anticipated by the inert gas rule by 2[97]. Of the 6 linear combinations derived from the acetylene π-functions one of a_2 (F^{π}_{3c}) symmetry finds no match with the metal orbitals and consequently one fewer antibonding orbital is produced than anticipated. Hoffmann and his coworkers have described some cyclo-butadiene complexes which show similar effects[98].

Acknowledgements. The S.E.R.C. is thanked for financial support and Prof. J. L. Templeton and R. L. Johnston for many helpful comments. The ideas presented in this review originated in 1981 in Geneva, where one of us (DMPM) was kindly invited to give a series of lectures on "Theory in Inorganic Chemistry". He thanks Dr. A. A. Williams for arranging this visit and many stimulating lunchtime discussions.

J. Appendix

Let $(\phi_1 \ldots\ldots \phi_N)$ be the set of N hydrogen 1s orbitals in the *closo* complex MH_N, and ψ_i^{closo} the i-th symmetry adapted linear combination of these orbitals:

$$\psi_i^{closo} = \sum_{j=1}^{N} C_{ij}\phi_j \qquad \text{AI}$$

This may be represented in matrix form as below:

$$\begin{pmatrix} C_{11} & C_{12} & \cdots & C_{1N} \\ C_{21} & C_{22} & \cdots & C_{2N} \\ \vdots & \vdots & & \vdots \\ C_{N1} & C_{N2} & \cdots & C_{NN} \end{pmatrix} \begin{pmatrix} \phi_1 \\ \phi_2 \\ \vdots \\ \phi_N \end{pmatrix} = \begin{pmatrix} \psi_1^{closo} \\ \psi_2^{closo} \\ \vdots \\ \psi_N^{closo} \end{pmatrix} \qquad \text{AII}$$

i.e.

$$\boldsymbol{C\Phi} = \psi^{closo} \tag{AIII}$$

Multiplication of Eq. AIII from the left by the inverse of the matrix of coefficients gives:

$$\boldsymbol{\Phi} = \boldsymbol{C}^{-1}\psi^{closo} \tag{AIV}$$

If interaction between the ligands is neglected, then:

$$\langle \phi_j | \phi_k \rangle = \delta_{jk}$$

and $\boldsymbol{C}$ becomes an orthogonal matrix whose inverse is equal to its transpose:

$$\begin{pmatrix} \phi_1 \\ \phi_2 \\ \cdot \\ \cdot \\ \phi_N \end{pmatrix} = \begin{pmatrix} C_{11} & C_{21} & \ldots & C_{N1} \\ C_{12} & C_{22} & \ldots & C_{N2} \\ \vdots & \vdots & & \vdots \\ C_{1N} & C_{NN} & \ldots & C_{NN} \end{pmatrix} \begin{pmatrix} \psi_1^{closo} \\ \psi_2^{closo} \\ \cdot \\ \cdot \\ \psi_N^{closo} \end{pmatrix} \tag{AV}$$

Thus to construct a hybrid from the set of ψ_i^{closo} functions which is localised exclusively on the j-th ligand, the following relationship is used:

$$\phi_j = C_{1j}\psi_1^{closo} + C_{2j}\psi_2^{closo} + \ldots\ldots C_{Nj}\psi_N^{closo} \tag{AVI}$$

$$\phi_j = \sum_{i=1}^{N} C_{ij}\psi_i^{closo} \tag{AVII}$$

The character of each *closo* symmetry adapted linear combination, $X_j(\psi_i^{closo})$, in this hybrid is simply the square of its coefficient:

$$X_j(\psi_i^{closo}) = C_{ij}^2 \tag{AVIII}$$

The character left on the *nido* fragment, ${}^{nido}X_j(\psi_i^{closo})$, is thus given by:

$${}^{nido}X_j(\psi_i^{closo}) = 1 - C_{ij}^2 \tag{AIX}$$

Construction of two hybrids localised exclusively over two ligands (applicable to consideration of the ligand orbital linear combinations in an *arachno* fragment) may be achieved by taking the following linear combinations:

$$\psi^+ = 1/\sqrt{2}\,(\phi_i + \phi_j) \tag{AX}$$

$$\psi^- = 1/\sqrt{2}\,(\phi_i - \phi_j) \tag{AXI}$$

Where ϕ_i and ϕ_j are expanded in accordance with Eq. AVII.

K. References

1. Lewis, G. N.: Valence and the Structure of Atoms and Molecules, Chemical Catalogue Co., New York, 1923; reprinted by Dover publications, New York 1966
2. Sidgwick, N. V.: Electronic Theory and Valence, Clarendon Press, Oxford 1927
3. Pauling, L.: J. Am. Chem. Soc. *53*, 1367 (1931)
4. Pauling, L.: Nature of the Chemical Bond, 3rd Ed., Cornell University Press, Ithaca, New York 1960
5. Richardson, J. W.: Organometallic Chemistry (Ed. H. Zeiss), Reinhold, New York, 12 (1960)
6. Sidgwick, N. V., Powell, H. E.: Proc. Roy. Soc., London, A 176, 153 (1940)
7. Gillespie, R. J., Nyholm, R. S.: Quart. Rev. *11*, 339 (1957)
8. Gillespie, R. J.: Molecular Geometry, Van Nostrand-Reinhold, Princeton, New Jersey 1972
9. Bartell, L. S., Plato, V.: J. Am. Chem. Soc. *95*, 3097 (1973)
10. Hall, M. B.: ibid. *100*, 6333 (1978), Inorg. Chem. *17*, 2261 (1978)
11. Walsh, A. D.: J. Chem. Soc., 2260; 2266; 2288; 2296; 2301; 2306; 2318; 2321; 2325; 2330 (1953)
12. Gimarc, B. M.: Molecular Structure and Bonding, Academic Press, New York 1979
13. Hoffmann, R.: Science *211*, 995 (1981)
14. Mehrotra, P. K., Hoffmann, R.: Theor. Chim. Acta *48*, 301 (1978) and references therein
15. Takahata, Y., Schnuelle, G. W., Parr, R. G.: J. Am. Chem. Soc. *93*, 784 (1971)
16. Burdett, J. K.: Struct. Bond. *31*, 67 (1976)
17. Hoffmann, D. K., Ruedenberg, K., Verkade, J. G.: ibid. *33*, 57 (1977)
18. Stone, A. J.: Mol. Phys. *41*, 1339 (1980)
19. Quin, C. M., McKieman, J. G., Raymond, D. B.: J. Chem. Educ. *61*, 569, 572 (1984)
20. Figgis, B. N.: Introduction to Ligand Fields, John Wiley and Sons, New York 1975
21. Klee, V.: Am. Scientist *59*, 81 (1971)
22. Fejes-Toth, H.: Regular Figures, Pergamon Press, Oxford 1964
23. Forsyth, M. I., Mingos, D. M. P.: J. Chem. Soc. Dalton Trans. 610 (1977) gives a general analysis of the capping problem
24. Mackay, A. L., Finney, J. L., Gotsh, K.: Acta Crystallogr. *33 A*, 98 (1977)
25. Craig, D. P., Doggett, G.: J. Chem. Soc. 4189 (1963)
26. Tolman, C. A.: Chem. Soc. Rev. *1*, 337 (1972)
27. Mitchell, P. R., Parish, R. V.: J. Che. Educ. *46*, 811, (1969)
28. Cotton, F. A., Wilkinson, G.: Advanced Inorganic Chemistry, Interscience, New York 1980
29. Green, M. L. H.: Organometallic Compounds, Vol. 2, Methuen, London 1968
30. Collman, J. P., Hegedus, L. S.: Principles and Applications of Organotransition Metal Chemistry, University Science Books, Mill-Valley, California 1980
31. Hoffmann, R., Adler, R. W., Wilcox, C. F.: J. Amer. Chem. Soc. *92*, 4992 (1972) gives a m.o. description of how square-planar carbon could be stabilised
32. Streitwieser, A., Owens, P. H.: Orbital and Electron Density Diagrams, MacMillan, New York 1973
33. Ginsberg, A. P.: Transition Metal Hydrides, Adv. in Chemistry Series *167*, 201 (1978)
34. Abrahams, S. C., Ginsberg, A. P., Knox, K.: Inorg. Chem. *3*, 538 (1964)
35. Burdett, J. K., Hoffmann, R., Fay, R. C.: ibid. *17*, 2553 (1978)
36. Burdett, J. K., Albright, T. A.: ibid. *18*, 2112 (1979)
37. Kepert, D. L.: Prog. Inorg. Chem. *24*, 179 (1978)
38. Hoffmann, R., Beier, B. F., Muetterties, E. L., Rossi, A. R.: Inorg. Chem. *16*, 511 (1977)
39. Kepert, D. L.: Prog. Inorg. Chem. *25*, 41 (1979)
40. Hoffmann, R., Howell, J. M., Rossi, A. R.: J. Amer. Chem. Soc. *98*, 2484 (1976)
41. Stiefel, E. I., Eisenberg, R., Rosenberg, R. C., Gray, H. B.: J. Am. Chem. Soc. *88*, 2956 (1966)
42. Keppert, D. L.: Prog. Inorg. Chem. *23*, 1 (1977)
43. Holmes, R. R.: ibid. *32*, 119 (1984)
44. Ballhausan, C. J., Gray, H. B.: Molecular Orbital Theory, W. A. Benjamin, New York 1964
45. Nyholm, R. S.: Proc. Chem. Soc. 273 (1961)
46. Hurley, A. C., Lennerd-Jones, J. E., Pople, J. A.: Proc. Roy. Soc. *A 220*, 446 (1953)
47. Hall, G. G., Lennard-Jones, J. E.: ibid. *A 202*, 155 (1950)

48. Pitzer, R. M.: J. Chem. Phys. *46*, 4871 (1971)
49. Murrell, J. N., Kettle, S. F. A., Tedder, J. M.: The Chemical Bond, John Wiley & Sons, New York 1978
50. Pauling, L.: Proc. Natl. Acad. Sci. (USA), 73, 274, 1043, 4290 (1976); 74, 2614, 5235 (1977); 75, 12, 569 (1978)
51. Kimball, G. E.: J. Chem. Phys. *8*, 188 (1940)
52. Doggett, G.: The Electronic Structure of Molecules: Theory and Application to Inorganic Molecules, Pergamon Press, Oxford 1972
53. Musher, J. I.: Angew. Chem., Intern. Ed. *8*, 54 (1969)
54. Muetterties, E. L.: Acc. Chem. Res. *3*, 266 (1970)
55. Hoffmann, R., Howell, J. M., Muetterties, E. L.: J. Am. Chem. Soc. *94*, 3047 (1972)
56. Rundle, R. E.: ibid. *85*, 112 (1963)
57. Rundle, R. E.: Surv. Prog. Chem. *i*, 81 (1963)
58. Williams, R. E.: Adv. Inorg. Chem. Radiochem. *18*, 67 (1976)
59. Albright, T. A.: Tetrahedron *38*, 1339 (1982) gives a detailed account of perturbation theory as applied to octahedral complexes
60. Cotton, F. A.: Chemical Applications Group Theory, Wiley Interscience, New York, 2nd ed. 1971
61. Steudal, R.: Chemistry of the Non-Metal, de Gruyter, Berlin, New York 1977
62. Musher, J. I.: Sulfur Research Trends, Advances in Chemistry Series (Ed. Gould, R. F.) American Chemical Society, Washington *110*, 44 (1972) and references therein
63. Pauling, L., Herman, Z. S.: J. Chem. Educ. *61*, 582 (1972)
64. Bent, H. A.: Chem. Rev. *61*, 275 (1961)
65. Mingos, D. M. P.: Comprehensive Organometallic Chemistry (Ed. Wilkinson, G., Stone, F. G. A., Abel, E. W.) *3*, 1 (1982)
66. Elian, M., Hoffmann, R.: Inorg. Chem. *14*, 1058 (1975)
67. Yared, Y. W., Miles, S. L., Bau, R., Reed, C. A.: J. Am. Chem. Soc. *99*, 7076 (1977)
68. Burdett, J. K.: Adv. Inorg. Chem. Radiochem. *21*, 113 (1978)
69. Burdett, J. K., Grzybowski, J. M., Perutz, R. N., Poliakoff, M., Turner, J. T., Turner, R. F.: Inorg. Chem. *17*, 147 (1978)
70. Burdett, J. K.: Coord. Chem. Rev. *27*, 1 (1978)
71. Simpson, M. B., Poliakoff, M., Turner, J. J., Maier II, J. B. McLaughlin, J. G.: J. Chem. Soc. Chem. Commun. 1355 (1983)
72. Elian, M., Chen, M. M. L., Mingos, D. M. P., Hoffmann, R.: Inorg. Chem. *15*, 1148 (1976)
73. Hoffmann, R.: Angew. Chem. Inter. Ed. 21, 711 (1982)
74. Mingos, D. M. P.: Acc. Chem. Res. *17*, 311 (1984)
75. Chatt, J., Leigh, G. J., Mingos, D. M. P.: J. Chem. Soc. Dalton, 1674 (1969)
76. Kubacek, P., Hoffmann, R.: J. Am. Chem. Soc. *103,* 4320 (1981)
77. Chisholm, M. H., Hoffmann, J. C., Kelly, R.: ibid. *101*, 7615 (1979)
78. Templeton, J. L., Ward, B. C.: ibid. 102, 6568 (1980)
79. Graham, W. A. G.: Inorg. Chem. *7*, 315 (1968)
80. Mingos, D. M. P.: J. Organometal. Chem. *179*, C29 (1979)
81. Kappe, T., Ziegler, E.: Angew. Chem. Inter. Ed. *13*, 491 (1974)
82. Jorgensen, W. L., Salem, L.: The Organic Chemist's Book of Orbitals, Academic Press, New York 1973
83. Schleyer, P. V. R.: New Horizons in Quantum Chemistry (Ed. Lowdin, P.-O., Pullman, B.) Reidel Dordrecht, 95 (1983)
84. Schleyer, P. V. R., Wurthwein, E.-U., Pople, J. A.: J. Am. Chem. Soc. *104,* 5839 (1982)
85. Wurthwein, E.-U., Schleyer, P. V. R., Pople, J. A.: ibid. *106*, 6973 (1984)
86. Carrol, P. J., Titus, D. D.: J. Chem. Soc. Dalton, 824 (1977)
87. Gillard, R. D.: Rev. Port. Quim. *11*, 70 (1969)
88. Searcy, A. W.: J. Chem. Phys. *28*, 1237 (1958)
89. San Filippo, J., Grayson, R. L., Snidoch, H. J., Inorg. Chem. *15*, 269 (1976); *16*, 1016 (1977)
90. Griffith, W. P., Pauson, D.: J. Chem, Soc. Dalton, 1315 (1973)
91. Uemura, S., Spencer, A., Wilkinson, G., J. Chem. Dalton, 2565 (1973)
92. Moore, D. S., Robinson, S. D.: Chem. Soc. Rev. *12*, 415 (1983)
93. Simons, J. H.: Fluorine Chemistry, Academic Press, New York, 1964, 5, 1

94. Fischer, E. O., Schubert, U.: J. Organometal. Chem. *100*, 59 (1975)
95. Mingos, D. M. P.: J. Chem. Soc. Dalton, 133 (1974)
96. Wooley, R. G.: Nouv. J. de Chemie *5*, 441 (1981)
97. King, R. B.: Inorg. Chem. *7*, 1044 (1968)
98. Chu, S. Y., Hoffmann, R.: J. Phys. Chem. *86*, 1289 (1982)

Molecular Mechanics and the Structure Hypothesis

Jan C. A. Boeyens

Department of Chemistry, University of the Witwatersrand, 1 Jan Smuts Avenue, Johannesburg 2001, South Africa

Chemical theories are so firmly based on the molecular-structure hypothesis that a reversal could be chaotic, without necessarily promoting comprehension. When dealing with large molecules there furthermore is a real need to define a rigid molecular structure and, rather than reject the structure hypothesis, it needs more precise exposition. Structures emerge by abstraction according to the Born-Oppenheimer prescription, which demands a commutative classical algebra of structural observables. Electron density can be modelled either quantum-mechanically or classically by the Hellmann-Feynman procedure. Molecular mechanics provides the classical framework for structural theory by incorporating quantum properties of the electron-density function in the form of a valence force field mapped onto the classical structure. This procedure clearly distinguishes between electronic and steric factors. Simulation of bond formation and rupture calls for a force field of anharmonic interactions. Molecular mechanical bonds are localized and cannot be parametrized in terms of molecular-orbital concepts. Bond orbitals to describe the bonding interaction in terms of electron pairs and positive cores are required. This level of abstraction is not incompatible with a point-charge model of the electron density as a function of internuclear distance. The abstraction inevitably introduces errors. The adiabatic approximation breaks down at, or near, symmetrical nuclear arrangements where electronic and vibrational functions are not clearly separable, causing Jahn-Teller-type distortion of the classical structure by quantum effects. The inverse effect is not that well defined or understood; but in the absence of an alternative explanation, it is offered as a possible source of error and a topic for future study in the interpretation of experimentally measured electron densities that yield negative deformation densities in bonds between valence-electron-rich atoms.

"There is an analogy between complex nuclei and complex molecules, but there is a fundamental difference too: for molecules the first principles from which we cannot calculate their properties are far better known."[1)]

Hendrik Casimir

I. Introduction

Chemical theories represent a curious blend of classical and non-classical, or quantum, concepts. The integration has been so thorough that the dualism is often overlooked or even denied. As a result components of the two essentially incompatible schemes are often mixed indiscriminately into non-sensical models. An elementary example is provided by the efforts of the novice to rationalize the shapes of molecules, invoking not only the quantum-mechanical ideas of orbital hybridization, but, at the same time, also the classical Valence-Shell-Electron Pair-Repulsion model[2)]. It creates an impossible tangle, until it is realized that the two methods represent totally independent approaches to the problem. Whereas VSEPR predicts the shapes of simple molecules, hybridization schemes only serve to rationalize the electronic distribution in a molecule of known structure.

All efforts to calculate empirical or classical quantities by the methods of quantum mechanics, at whatever level of sophistication, suffers from the same weakness. To achieve a synthesis of empirical and quantum-theoretical concepts, it is necessary to develop the appropriate formalism for a theory of classical entities in parallel with the quantum theory, which is commonly used in chemistry only to describe electronic behaviour.

A possible alternative is to discard all classical or empirical concepts as redundant, on the assumption that any observable molecular property can in principle be derived from the quantum-mechanical model. The purpose of this review is to argue that this would be detrimental and that chemical theory relies equally on classical and on quantum-mechanical concepts. Molecular structure as the most basic of classical concepts[3)] especially deserves an appropriate theoretical description.

1. Historical Summary

Early use of the molecular hypothesis that originated with Avogadro[4)] made no or little distinction between different molecules or even atoms on the basis of size or shape. The concept of a molecular structure is generally thought to have started with Kekule[5)] and was gradually refined during the nineteenth century through the recognition of isomerism in organic chemistry[6)]. The first successful structural hypothesis was advanced by Van't Hoff[7)] who demonstrated that the optical activity of carbon compounds could be explained in terms of tetrahedral fragments imparting chirality, as earlier inferred by Pasteur[8)] to individual molecules and hence to the corresponding substance.

It is noted that Le Bel[9)] reached a similar conclusion, but considered the problem as one of symmetry only, without introducing the concept of a rigid molecular structure. The idea of a molecular structure, however, found such ready application in conceptualizing chemical change that the Van't Hoff model gained universal acceptance. It relates optical activity to molecular chirality, which is a geometrical property that clearly defines the dimensionality of molecular space[10]. It also introduced the notion of bonds between atoms in a molecule.

Application of these ideas revolutionalized chemistry during the four decades between Van't Hoff and Bohr[11]. At the time when quantum ideas were first introduced into chemistry[11] all current theories were formulated firmly on the basis of the molecular structure hypothesis. It is not surprising therefore to find that the early applications of quantum theory to chemistry consisted of grafting the new concepts onto existing structural models. The resulting hybrid theories persist to this day. Woolley[12] cites the following examples: the separation of molecular energies into electronic, vibrational and rotational contributions[13, 14] which is crucial for analysis of molecular structure in terms of bond-lengths, bond-angles and moments of inertia; the Franck-Condon principle[15, 16] and potential energy surfaces and the angular-momentum classification of molecular electronic orbits of diatomic molecules[17, 18]. It is also noted that the quantum-mechanical valence-bond method of Heitler and London[19], the starting point of the quantum theory of chemical bonding[20], likewise depends on the idea of fixed nuclei to provide a framework for the analysis of electronic distribution. In general the concept of a chemical bond, best defined in terms of a potential energy curve[21] as shown in Fig. 1, has no meaning outside the model that effectively separates nuclei and electrons, as described by the Born-Oppenheimer approximation[22, 23].

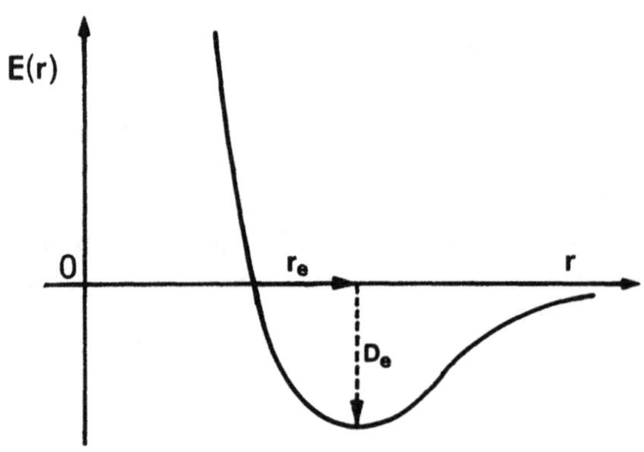

Fig. 1. An isolated chemical bond represented as a minimum (D_e) in the potential energy of interaction between a pair of atoms at an equilibrium separation r_e

Further examples of chemically fundamental concepts, as yet undefined quantum-mechanically, can readily be added to this list. It includes chirality, optical activity, isomerism[24], steric strain, torsion, barrier to rotation, crystal packing forces, force constant, bond energy, atomic polarization, electrostatic inductive effects, and all other concepts associated with molecular structure or chemical bonding. Even the *ab initio* methods of quantum chemistry follow the procedure of empirically assuming molecular structure as a first principle.

2. Quantum Theory and Structure

Woolley[12] examined the possibility of quantum-mechanically formulating the notion of molecular structure and found it to be an irreducible idea. Quantum theory derives the properties of complex systems deductively from the equations of motion of elementary particles, and one looks for molecular structure to appear as a derived concept. Since this has never been achieved and probably is inherently impossible, it follows that the concept of structure is either quantum-mechanically undefined[25] or created in condensed media as a result of the strong intermolecular interactions in the many-body environment[26].

The problem originates from the fact that, in principle, all electrons and nuclei forming a molecule should be consistently treated quantum-mechanically, i.e. they should be described by a Complete Molecular Eigenstate[3] (CMES), viz. an eigenstate of the complete molecular Hamiltonian, involving both electrons and nuclei:

$$H = T_N + T_e + V \tag{1}$$

where the nuclear and electronic kinetic energies are the sums of one-particle operators:

$$T_N = \sum_{\alpha=1}^{N} \left(-\frac{h^2}{8\,\pi^2 M_\alpha} \nabla^2_{R_\alpha} \right) = \sum_{\alpha} \frac{P^2_\alpha}{2\,M_\alpha} \tag{2}$$

$$T_e = \sum_{i=1}^{E} \left(-\frac{h^2}{8\,\pi^2 m_i} \nabla^2_{r_i} \right) = \sum_{i} \frac{p_i^2}{2\,m_i} \tag{3}$$

R_α and r_i are nuclear and electronic coordinates with respect to the centre of mass.

The potential energy V consists of the sum of the Coulomb interactions between all pairs of particles:

$$V = -\sum_{i,j=1}^{N,E} \frac{(Z_j e^2)}{|r_i - R_j|} + \sum_{\substack{i,k=1 \\ i>k}}^{E} \frac{(e^2)}{|r_i - r_j|} + \sum_{\substack{j,l=1 \\ j>l}}^{N} \frac{(Z_j Z_l e^2)}{|R_j - R_l|} \,. \tag{4}$$

There is no distinction between electrons and nuclei and no mechanism whereby relative atomic positions can be localized.

Due to the symmetry properties of this Hamiltonian, reflected in the CMES, i.e. invariance with respect to all rotations, a non-degenerate CMES should, in principle, exhibit full spherical symmetry for any molecule. The common assumption that each eigenstate corresponds to a definite molecular structure is thus not justified[3].

A quantum-mechanical definition of structure was considered in more detail by Claverie and Diner[27]. They define "quantal" or "potential" structure in the sense that certain spatial correlations between nuclei can be computed from the CMES. The simplest example is provided by a diatomic molecule: for any given CMES one may compute the probability density P(R) for the internuclear distance R, and as a general rule this density will peak around some "equilibrium distance", R_0. Such properties may reflect some underlying structure. In fact, the same type of correlation should also exist between electrons. This notion of structure is a quantum-mechanical concept and should not be confused with the much stronger idea of a classical molecular structure, which is clearly quantum-mechanically undefined. The essential difference is that the quantal structure manifests itself only when some measurement takes place while classically a molecule could be considered as a set of point-like nuclei endowed with a well-defined position at every instant of time, independent of any measurement[3].

The notion of a spatial structure for polyatomic molecules has physical meaning only in terms of the Born-Oppenheimer[22] approximation: For a set of nuclei and electrons with coordinates Q_α and q_i, respectively, the Schrödinger equation is

$$H\Psi(Q,q) = \left[\sum_\alpha \frac{P_\alpha^2}{2M} + \sum_i \frac{p_i^2}{2m} + V(Q,q)\right]\Psi(Q,q) = E\Psi(Q,q) \tag{5}$$

with $V(Q,q)$ as defined in (4) above. Clamping the nuclei into fixed positions Q' eliminates the nuclear kinetic energy, and the electronic wave equation with eigenvalues E' becomes

$$\left[\sum_i \frac{p_i^2}{2m} + V(Q',q)\right]\psi(Q',q) = E'(Q')\psi(Q',q)\ . \tag{6}$$

This equation can in principle be solved for all conceivable nuclear configurations. In any particular case it is assumed that

$$\Psi(Q,q) = \psi(Q,q)v(Q) \tag{7}$$

where v is some vibrational function. Substitution into (5) gives

$$\begin{aligned} H\psi(Q,q)v(Q) &= E\psi(Q,q)v(Q) \\ &= \left[\sum_\alpha \frac{P_\alpha^2}{2M} + \sum_i \frac{p_i^2}{2m} + V(Q,q)\right]\psi(Q,q)v(Q)\ . \end{aligned} \tag{8}$$

The nuclear kinetic energy is

$$\begin{aligned} \frac{P^2}{2M}\psi(Q,q)v(Q) &= -\frac{h^2}{8\pi^2 M}\nabla_\alpha^2\psi(Q,q)v(Q) \\ &= -\frac{h^2}{8\pi^2 M}\left[\psi(Q,q)\nabla_\alpha^2 v(Q) + \nabla_\alpha\psi(Q,q)\cdot\nabla_\alpha\psi(Q) + v(Q)\nabla_\alpha^2\psi(Q,q)\right]\ . \end{aligned} \tag{9}$$

Since the nuclei are massive compared to electrons, $m/M \ll 1$, the electronic wavefunction is assumed to adjust to changes in nuclear position or momentum, and (9) reduces to

$$\begin{aligned} \frac{P^2}{2M}\psi(Q,q)v(Q) &\simeq -\frac{h^2}{8\pi^2 M}\psi(Q,q)\nabla_\alpha^2 v(Q) \\ &= \psi(Q,q)\frac{P^2}{2M}v(Q)\ . \end{aligned} \tag{10}$$

For any set of nuclear positions, a stationary distribution of electrons, described by a wavefunction rather than a density matrix is therefore assumed. This means that, in the limit of infinite nuclear mass, electrons follow the nuclei adiabatically[28)].

Substitution of this result into (8) yields

$$\sum_{\alpha} \psi(Q,q)\frac{P_{\alpha}^{2}}{2M}\nu(Q) + \left[\sum_{i}\frac{p_{i}^{2}}{2m} + V(Q,q)\right]\psi(Q,q)\nu(Q) = E\psi(Q,q)\nu(Q)\,. \qquad (11)$$

According to (6), the second term on the left becomes $E'\psi(Q,q)\nu Q)$ and (11) simplifies to

$$\left[\sum_{\alpha}\frac{P_{\alpha}^{2}}{2M} + E'(Q)\right]\nu(Q) = E\nu(Q) \qquad (12)$$

which shows that the wavefunction for the nuclear motion is calculated in an effective potential $E'(Q)$ obtained from the energy eigenvalues of the stationary electronic state. It is noted that without the adiabatic approximation the concept of molecular structure has no physical meaning.

3. *Small and Large Molecules*

For small, isolated molecules one finds that their physical properties are not modeled well in terms of a rigid molecular structure and that the full quantum-mechanical description of the system is needed in order to account for all experimental observations[29)]. The inversion of ammonia[30)] presents the earliest example of a molecule not properly described in terms of a rigid classical structure. Another familiar example is cyclopentane[31)] and all other molecules that exhibit pseudo-rotation. Techniques like laser spectroscopy, particle scattering, and molecular beams reveal numerous examples[32)] of molecular non-rigidity and electron-nuclear (vibronic) coupling not adequately described in terms of the traditional Born-Oppenheimer approximation[22)].

For larger molecules, however, or for molecules in the solid state this procedure is neither practical, due to its complexity, nor theoretically justified; for these systems, molecular rigidity and a robust structure are experimentally observable, e.g. by diffraction analysis. The first point is well illustrated by the example of boromycin cited by Prelog[10)]. It has 18 chiral centres and hence 2^{18} possible stereoisomers, only one of which occurs naturally. To identify the unique isomer in terms of a three-dimensional structure is an almost trivial operation compared to a quantum-mechanical description that would treat the 262,144 possibilities, and more, simultaneously within a single molecular Hamiltonian. As boromycin represents a relatively simple arrangement in comparison with proteins and other biological molecules, it provides compelling reasons why the structure hypothesis is justified, even if only for practical reasons, for the important class of larger molecules. The traditional superposition of the structure concept on quantum theory, however, is of questionable validity. To achieve the proper synthesis of ideas presents a mathematical problem which has been formulated in detail by Primas[33, 34)]. The relevant arguments are briefly summarized in the next section.

II. Entangled Systems

It seems clear that chemical theory, for practical reasons, requires the definition of classical entities in addition to the quantum-mechanical. The problem is to establish the relationship between the two classes of object and their mode of entanglement. The definition of a molecule then reduces to finding the appropriate abstraction from entangled reality to represent an isolated classical object with structure. This, of necessity, ignores all interaction between the classical and non-classical parts of the same entity. Quantum mechanical distortion of classical structures can therefore be anticipated and, as shown in Sec. V, it is a general feature of many structures. The inverse effects are of more speculative reality.

The theory is based on an observation of Einstein, Podolsky, and Rosen[35] (EPR) that previously interacting quantum-mechanical systems remain correlated afterwards even if the interaction does not persist. This is inferred from the demonstration that, in the absence of such correlation, a paradox occurs for observable properties characterized by non-commuting operators; i.e.

$$AB \neq BA$$

whereby precise knowledge of property A precludes all knowledge of B.

Consider two systems (I and II) which interact during the period $0 \leq t \leq T$, after which there is no further interaction. Suppose at $t < 0$ the states of both systems were known. The state of the combined system at any $t > 0$ can be calculated, in particular at $t > T$. However, one cannot calculate the state of any one of the systems after the interaction. According to quantum mechanics this can only be done by further measurement in a process known as reduction of the wave packet.

One assumes eigenvalues $a_1, a_2 \ldots$ of property A pertaining to system I, with corresponding eigenfunctions $U_1(x_1), U_2(x_1), \ldots$ etc., where x_1 represents the variables used to describe system I. Then at $t > T$

$$\psi(x_1, x_2) = \sum_{n=1}^{\infty} \psi_n(x_2)U_n(x_1) \tag{13}$$

where x_2 are the variables used to describe system II.

Now suppose A is measured at value a_k. The measurement leaves system I in state $U_k(x_1)$, the second system in state $\psi_k(x_2)$, and the wave packet is reduced to $\psi_k(x_2)U_k(x_1)$. The set of functions $U_k(x_1)$ is determined by the choice to measure quantity A. An alternative would be to measure property B, with eigenvalues $b_1, b_2 \ldots$ etc. and eigenfunctions $V_1(x_1), V_2(x_1), \ldots$ etc. At $t > T$ one then writes

$$\psi(x_1, x_2) = \sum_{s=1}^{\infty} \phi_s(x_2)V_s(x_1) \tag{14}$$

and measurement of B gives the eigenvalue B_r. This leaves the wave packet

$$\phi_r(x_2)V_r(x_1) .$$

It follows that two different measurements on system I leaves system II in states with two different wave functions, $\psi_k(x_2)$ and $\phi_r(x_2)$; and since the two systems no longer interact, no real change in system II should be brought about by anything done to system I.

It may be that the wave functions are eigenfunctions of two non-commuting operators corresponding to some physical quantities P and Q respectively. Then, by measuring either A or B in system I, it is possible to predict with certainty and without disturbing the second system either the value of $P(p_k)$ or of $Q(q_r)$. In the first case P is an element of reality and in the second case Q is an element of reality. But ψ_k and ϕ_r belong to the same reality. This contradicts the previous conclusion that non-commuting operators cannot have simultaneous reality.

The nature of the paradox is put into perspective by Bell's theorem[36], discussed in detail by Clauser and Shimony[37] and also d'Espagnat[38, 39]. The paradox only exists in the context of realistic local theories, which assume that physical reality exists independent of human observation, that inductive inference is a valid mode of reasoning, and since no influence can propagate faster than the speed of light, that action at a distance is impossible. The conditions of realism and of Einstein separability are not implicit in quantum theory. It follows that quantum theory, unlike realistic local theory, is not inconsistent with the correlation of distant events. The Einstein paradox therefore is amenable to analysis by experiments set up to distinguish between the predictions of quantum theory and Bell's theorem, also known as Bell's inequality. The most popular type of experiment concerns singlet systems; i.e. spin 0 systems made up of particles with spin $\pm 1/2$. A pair of protons produced by the decomposition of a hydrogen molecule represents such a system. The experimental details for this and other tests are reviewed by Clauser and Shimony[37]. In all experiments there is overwhelming support for the predictions of the quantum theory, establishing the correlation of distant events as experimental fact.

The conclusion is that the physical world cannot be correctly described by a realistic local theory. It is necessary either to abandon the criterion of reality or to accept some kind of action-at-a-distance. Since both options are radical, theorists are reluctant to choose; but d'Espagnat[39] seems to favour the latter. He argues that the wave function for a system of two or more particles is in general a non-local entity which collapses when a measurement is made. This collapse is instantaneous and complete. It occurs everywhere, also at the position of the particle not involved in the measurement. Measurements of system I can therefore affect system II via the collapsing wave packet. In terms of quantum theory no paradox therefore exists, and the correlation of distant events is predicted.

Most particles or aggregates of particles usually regarded as separate objects have interacted in the past with other objects and must hence remain correlated to constitute an indivisible entangled whole. The concept of reality in such a world must take these correlations into account and therefore has a different meaning from the reality of everday common experience. It follows that the quantum world is holistic, and one senses that this would complicate efforts to define an isolated object. It can only be done by ignoring the EPR correlations between an object and its environment, which clearly is not a trivial approximation.

1. Emerging Molecules

It now becomes apparent that whereas few chemists would doubt the existence of identifyable molecules, finding a scientifically acceptable description of an isolated molecule is non-trivial. A molecule exists only if it can be distinguished from its environment. Consequently, basic to the science of chemistry is the assumption that molecular systems can be separated from the environment with which they are entangled and that this could be achieved without violating the dictates of quantum theory.

To illustrate the process of abstraction, Primas[34] gives the following example:

According to quantum-mechanics the electrons of the moon are entangled with their radiation field. Unless an abstraction is made from this radiation field, the moon becomes entangled with the sun and cannot be said to possess an individuality. So, without abstracting from the quantum structure of the radiation field, the moon cannot be an object.

Every description of nature depends on the division of the world into a part whose effects are to be considered, and another part whose effects can be ignored, i.e. into essential and accidental components. The essential part is not necessarily an intrinsic property, but rather depends on the particular point of view adopted. A point of view is characterized by a deliberate lack of interest in what breaks the holistic unity of nature. Every description of nature presupposes a particular point of view. Observable phenomena come into being by breaking the holistic symmetry. Any alternative viewpoint with a different emphasis leads to an inequivalent description. There is one reality, but there are many points of view. These are called complementary descriptions, and no experiment can be devised that could demonstrate complementary aspects in a single observational context.

The quantum-mechanical description is complementary to the chemical description. Both describe the same system, but they have different viewpoints, hence they use different algebras of observables. The quantum-mechanical model does not allow the description of classical features, while the chemist describes a molecule as having a classical molecular structure.

In the quantum-mechanical description electrons and nuclei are entangled by EPR correlation. This is the appropriate model to use for the description of isolated, small flexible molecules where the supposition of a rigid structure is inappropriate[32]. In the chemical description of larger molecules, electrons and nuclei are considered as not correlated, in spite of their strong electromagnetic interaction, so that nuclei exist as classically describable individual objects. Formulation of a theory that recognizes aspects of both points of view is required for the description of a chemical system. This demands an extended logic that recognizes the holistic nature of reality.

2. *Quantum Logic*

According to the postulates of quantum mechanics, observables can be defined mathematically in terms of the properties of their corresponding operators, which constitute an algebraic measure space, i.e. a closed set of operational rules. The appropriate measure space that correctly describes quantum variables is called a Hilbert space, L_2. The space for classical observables is the measure space L_∞. Both of these belong to the general class of Banach spaces, and in particular they are both dual Banach spaces, also called W*-algebras. Whereas L_∞ has a commutative W*-algebra, that for L_2 is non-commutative. These are the algebras of observables for these systems.

The W*-algebra of quantum observables is necessarily non-commutative to allow for the co-existence of non-commuting operators, amounting to simultaneous definition of classically incompatible properties. This defines a non-Boolean quantum logic. Abstraction from this non-Boolean quantum world amounts to emphasizing only one of the incompatible properties, and hence this abstraction (projection) is correctly described in a Boolean (two-valued) frame of reference. This is made possible only by ignoring the surroundings and the non-Boolean correlation of the object with its environment.

The W*-quantum logic provides a comprehensive global description embedding local descriptions with Boolean frames of reference. In the same way a closed W*-system with a W*-algebra (A) of observables envelopes sub-systems with sub-algebras, so that A can, for instance, be written as a tensor product of two W*-algebras, B and C:

$$A = B \otimes C\,.$$

The basic theory is said to be factorized into two subtheories

$$T = T_1 \otimes T_2\,.$$

In an object factorization, one sub-system (T_1) is called the object and T_2 its environment.

According to Raggio's theorem[40], $T_1 \otimes T_2$ is an object factorization only if one of the subtheories is classical, i.e. has a commutative algebra. Accordingly only three kinds of object can exist:

(i) classical objects embedded in a classical environment;
(ii) classical objects embedded in a non-classical environment;
(iii) non-classical objects embedded in a classical environment.

Significantly the case of non-classical objects interacting with a non-classical environment is logically excluded.

Therein lies the root of the problem pertaining to the analysis of molecular structures by the conventional methods of quantum-mechanics. Unless a suitable abstraction is made, the electrons and nuclei are correlated in the EPR sense so that neither electrons nor nuclei exist as individual objects and the concept of structure has no meaning. In an abstraction that ignores these correlations, but not electromagnetic interactions, electrons in a molecule can be separated from the nuclei in the Born-Oppenheimer sense. However, in this abstraction both electrons and nuclei can no longer be described quantum-mechanically. At least one set needs to be described classically. The price of the Born-Oppenheimer approximation is a classical molecular structure.

III. Theoretical Study of Molecules

For practical reasons all molecules of any complexity are best described in terms of a robust molecular structure, generated by a separation of electronic and nuclear motions. The separation is only possible when ignoring EPR correlation between the electron cloud and the nuclear framework. In practice the electron cloud is commonly described by some approximate molecular wave function as a non-classical entity. The nuclear positions serve as boundary conditions for the interpretation of the wave function. The nuclear distribution, which constitutes the molecular structure, is therefore of vital importance in any electron-density analysis.

The molecular structure, on the contrary, is a classical coherent entity. The cohesion is provided by the charge cloud manifesting itself at the classical level as electrostatic interaction. This chemical bonding can therefore be described in any of three different ways:

(i) If the equilibrium positions of the nuclei are known, a static structure is assumed, and the interaction of the electronic charge cloud with the classical nuclear configuration is calculated by quantum-mechanical methods. Depending on computational detail, this procedure is known as either the molecular-orbital or the valence-bond method.

(ii) Since chemical bonding is a phenomenon strictly associated with a molecular structure and does not occur in a non-adiabatic quantum-mechanical analysis, it can be formulated correctly in purely classical terms. The problem reduces to classical electrostatics whereby the charge cloud is described in terms of an electron-density function, or a set of equivalent point charges, interacting with the positively charged nuclei, as prescribed by the Hellmann-Feynman theorem[41, 42].

(iii) Ignoring the physical origin of the cohesion allows the chemical bond to be most simply described in purely mechanical terms. The effect of the charge cloud is then summarized by a force field that specifies all interactions between possible pairs of atomic nuclei. This is the method of molecular mechanics.

1. Electronic and Steric Interactions

All conceptual schemes for the study of molecular structure rely on the separation of electronic and nuclear components and are therefore necessarily incomplete. In the quantum-mechanical description of an electronic charge distribution, the geometrical shape of the molecule is assumed to model all interactions through bonds, but not through space. In the classical stereochemical treatment, however, the molecular geometry is optimized in terms of the quantum-mechanical bonding pattern and non-bonded interactions. All parameters are hence never refined simultaneously as in the quantum-mechanical, non-adiabatic treatment of small, flexible molecules[32]. It is therefore logically expedient to formally distinguish between electronic and steric interactions, although there is really no fundamental difference between the two sets of interaction.

It follows that an experimentally observed molecular structure, which represents an equilibrium arrangement in which steric and electronic factors are balanced, cannot be modeled correctly in terms of either electronic or steric factors only. It is necessary to describe steric effects by classical methods (e.g. molecular mechanics) and electronic effects quantum mechanically (e.g. LCAO). However, the two procedures are not independent: molecular mechanics relies on a force field derived from the electronic distribution whereas LCAO requires the molecular structure as a starting point. For practical reasons, to be discussed presently, it is advisable first to eliminate steric complications by means of molecular mechanics before quantum-mechanical calculations commence.

2. *Molecular Mechanics*

As required by theory the most successful method for calculating molecular structure is a classical model, known as either molecular mechanics, the empirical-valence force-field method, or variations thereof. One of the many comprehensive reviews of the method was published comparatively recently by White[43].

In terms of molecular mechanics, a molecule is described as an interacting set of atoms. The equilibrium configuration is considered to correspond to a minimum in the molecular potential-energy function (Born-Oppenheimer surface), which takes into account interactions between all possible atomic pairs in the molecule. The zero point of the potential function is defined to correspond to an arrangement in which each chemical bond, valence angle, bend or torsion has an ideal value determined by electronic factors only. In most applications, the method only treats small distortions away from the electronic values; and to calculate the strain so introduced, harmonic restoring forces are assumed. The potential strain energy for the deformation of a parameter p from its strain-free value of p_0 is specified in terms of the harmonic restoring force

$$F = -\frac{\partial V}{\partial p} = k_p(p - p_0) \tag{15}$$

where the proportionality constant k is known as a force constant. Integration gives the potential energy

$$V(p) = k_p(p - p_0)^2 \tag{16}$$

since $V(p_0) \overset{\text{def}}{=} 0$. The total strain in a molecule is obtained as the sum over all interactions

$$V_t = \frac{1}{2}\sum_{i=1}^{N} k_i(\Delta p_i)^2 = V_l + V_\theta + V_\chi + V_\phi + V_r + V_q \tag{17}$$

including bond and angle deformation, bending and torsional terms, non-bonded and coulombic[44] interactions.

It is standard practice to assume harmonic potentials for the first two terms only and to treat the other by special methods. In principle a single type of anharmonic potential, like a Morse[45] function, could be used for all interactions, but this approach has not been generally accepted.

It is finally assumed that with all force constants and potential functions correctly specified in terms of the electronic configuration of the molecule, the nuclear arrangement that minimizes the steric strain corresponds to the observed molecular structure of the isolated (gas-phase) molecule. The object of the exercise is to minimize the intramolecular potential energy, or steric energy, as a function of the nuclear coordinates. The most popular procedure is by a computerized Newton-Raphson method. Most program libraries contain efficient subroutines to apply this alogrithm for virtually any number of terms. It works on the basis that the vector $\mathbf{V}_t'$ with elements $\partial V_t/\partial x_i$, the first partial derivatives with respect to cartesian coordinates, vanishes at a minimum point, i.e.

$$\mathbf{V}_t'(\mathbf{x}^*) = 0 \; . \tag{18}$$

This implies zero net force on each atom in the molecule. Let **x** represent a position vector representing some trial structure and let $\delta\mathbf{x}$ be the vector of required corrections. Then

$$\mathbf{x}^* = \mathbf{x} + \delta\mathbf{x} \; . \tag{19}$$

Hence

$\mathbf{V}_t'(\mathbf{x} + \delta\mathbf{x}) = 0$, which by Taylor expansion becomes

$$\mathbf{V}_t'(\mathbf{x} + \delta\mathbf{x}) = \mathbf{V}_t'(\mathbf{x}) + \mathbf{V}_t''(\mathbf{x})\delta\mathbf{x} + \ldots = 0 \tag{20}$$

For computational simplicity the series is usually truncated after the terms linear in $\delta\mathbf{x}$. Then

$$\mathbf{V}_t'(\mathbf{x}) + \mathbf{V}_t''(\mathbf{x})\delta\mathbf{x}_a = 0 \tag{21}$$

where $\delta\mathbf{x}_a$ is an approximate correction and $\mathbf{V}_t''(\mathbf{x})$ is the Hessian matrix of second derivatives $\partial^2 V_t/\partial x_i \cdot \partial x_j$. The correction vector is then written as

$$\delta\mathbf{x}_a = \beta\mathbf{V}_t''(\mathbf{x})^+\mathbf{V}_t(\mathbf{x}) \tag{22}$$

where β is a damping factor to prevent divergence at the early stages of refinement. $\mathbf{V}_t''(\mathbf{x})^+$ is the inverse of $\mathbf{V}_t''(\mathbf{x})$.

3. Force Fields

The potential-energy expressions used in molecular mechanics derive from the force-field concept that occurs in vibrational spectroscopic analysis according to the **GF**-matrix formalism[46,47)]. The **G**-matrix contains as elements atomic masses suitably reduced to match the internal displacement coordinates (matrix **D**) in defining the vibrational kinetic energy T of a molecule:

$$2T = \dot{\mathbf{D}}^T\mathbf{G}\dot{\mathbf{D}} \; .$$

In the same notation the potential energy is obtained as

$$2V = \mathbf{D}^T\mathbf{F}\mathbf{D}$$

where the matrix **F** has the quadratic force constants,

$$k_{ij} = \frac{\partial^2 V}{\partial D_i \partial D_j}$$

as elements. The classical equations of motion[1] for internal coordinates

$$\frac{d}{dt}\frac{\partial T}{\partial \dot{D}_i} + \frac{\partial V}{\partial D_i} = 0 \tag{23}$$

have amplitude solutions of the type

$$D_n = A_n \cos(\alpha t + \varepsilon)$$

provided the secular determinant of coefficients vanishes, i.e.

$$|\mathbf{F} - \lambda \mathbf{G}^{-1}| = 0 \quad \text{or} \quad |\mathbf{FG} - \lambda \mathbf{E}| = 0 \,. \tag{24}$$

The solutions $\lambda_i = \alpha_i^2 = 4\pi^2\nu_i^2$ yield the normal vibrational frequencies of the molecule.

It follows that force constants can in principle be derived from the observed vibrational frequencies, but in practice one finds an insufficient number of experimental frequencies to allow determination of all the force constants in the **F**-matrix. This seriously limits the use of vibrational spectra in the derivation of transferable force constants required for the general application of molecular mechanics.

The problem is partially overcome experimentally by the technique of isotopic substitution. For any non-linear molecule of N nuclei the number of frequencies is limited to 3N–6. Any change in atomic mass brought about by isotopic substitution results in a different spectrum and hence an increase of experimental information. It can be demonstrated in terms of product rules[48)] that the frequencies of the substituted and unsubstituted species are not independent and that the extra information is equivalent to only one force constant per isotopic substitution. The frequency data may also be augmented by the Coriolis coupling constants derived from the rotational fine structure of vibrational bands[49)]. For small molecules it might even be possible to measure more primary frequency data than the number of independent force constants and enable least-squares refinement. In other cases[50)] the frequency data of homologous compounds can be combined in a single calculation to improve the overdetermination.

The primary object, in the present context, is to obtain transferable force constants for bond stretching, angle deformation, bending, and torsion. These are the diagonal elements of the **F**-matrix. It appears that the more favourable the ratio of data to variables, the closer the calculated force constants approach the ideal of transferability. Aron et al.[51)] compared force constants of an extensive series of halogenated methanes, silanes and germanes. They demonstrated impressive correlation with several related properties like bond length and bond strength, implying a high degree of transferability. It appears that in chemically related, well refined systems the diagonal elements seem to converge to constant transferable values, whereas most of the variation accumulates off-diagonally. The physical meaning, at least of those force constants required for molecular mechanics, need therefore not be in doubt. Zerbi[52)] recommends an overdetermination of at least 4/3 in order to achieve sufficient realiability and is generally much more pessimistic about the physical meaning of quadratic force constants. Nevertheless he

1 The rate of change of momentum (first term) is equal to the force acting upon the mass (negative of the second term)

recognizes the positive aspects of force fields in structure analysis, vibrational assignment, lattice dynamics, phonon dispersion and the calculation of elastic constants.

Most of the force constants currently used in molecular mechanics are, however, not too well established experimentally. In general they derive from spectroscopic analyses based on approximate force fields, which correspond to **F**-matrices simplified by the eliminination of certain non-zero elements, or some other device. The Cotton-Kraihanzel[53] force field, developed especially for metal carbonyls, provides an interesting illustration of such a procedure. Their definition of mathematical relationships between off-diagonal interaction force constants, established on chemical grounds, reduced the number of terms sufficiently to allow the calculation of transferable stretching force constants.

Several simplified force fields of more general utility are in circulation, including the central force field (CFF), valence force field (VFF) and Urey-Bradley force field (UBFF). The central force field assumes that molecules are held together by forces acting along lines between all pairs of atoms, the valence force field describes the interaction mainly in terms of the diagonal force constants, and the Urey-Bradley force field replaces off-diagonal interaction constants by special non-bonded interactions[54].

The simplest of these[55] (CFF) treats all possible pairwise interactions between atoms in the molecule as simple harmonic vibrations

$$V = \tfrac{1}{2} \sum_{i>j}^{N} k_{ij}(\delta_{ij} - \delta_{ij}^{0})^2 .$$

Very little use is made of this approach in molecular mechanics although it provides a convenient method to assign conformations to molecular formulae[56]. Force constants clearly have very little physical meaning in this formalism.

Most force fields used in molecular mechanics are of the valence type and are derived from the generalized VFF, i.e. the complete **F**-matrix. It is known from experience that interaction constants of the valency type tend to be considerably smaller than the diagonal elements of **F**. Setting all of the off-diagonal elements to zero produces the Simple VFF. This amounts to the assumption that bonding interactions are sufficiently localized to confine the normal molecular vibrations to independent neighbourhoods spanning no more than three contiguous covalent bonds. Quantum-mechanically these interactions are described in terms of orthogonal molecular orbitals localized between two, three or four nuclei. In the classical model they are quantified in terms of bond stretching, angle bending, out-of-plane bending, and torsional force constants. It is generally conceded that this simplification is too drastic, and molecular mechanics requires an intermediate VFF with additional terms representing interaction between localized vibrations and between more remote centres. Mathematically this defines a potential

$$V = \tfrac{1}{2} \sum k_r(\delta - \delta_0)^2 + \tfrac{1}{2} \sum k_\theta(\theta - \theta_0)^2 + \tfrac{1}{2} \sum k_\chi(\chi - \chi_0)^2 + \tfrac{1}{2} \sum k_\phi(\phi - \phi_0)^2 + \\ + \sum\sum_{|i-j|<3} k_{ij}(p_i - p_i^0)(p_j - p_j^0) + \tfrac{1}{2} \sum_{|i-j|>2} (\delta_{ij}) . \quad (25)$$

Since bond torsions are often too large to justify the harmonic approximation, special potentials are used in molecular mechanics to describe distortions of this type.

The UBFF differs from the GVFF only in the treatment of off-diagonal elements[57]. Instead of interaction force constants, it introduces a completely new feature, namely, repulsion between non-bonded atoms. The repulsion diminishes as the distance between atoms increases. It can be shown[47] that for 1, 3 interaction between atoms bonded to the same central atom, the interaction constants (stretch-stretch and bend-stretch) are equivalent to repulsions provided linear terms are included in addition to the usual quadratic terms. The total potential energy is equivalent to the VFF potential defined before, but with the cross-terms replaced by:

$$\tfrac{1}{2}\sum k_{13}(r_{13} - r_{13}^0)^2 + \sum k_{13}'(r_{13} - r_{13}^0) + \sum k_r'(r - r_0) + \sum k_\theta'(\theta + \theta_0)\ .$$

Although the UBFF is often preferred in spectroscopic analysis, its use in molecular mechanics is rather limited especially since a SVFF often suffices.

4. Force Constants and Potential Functions

Exact molecular-mechanics simulation of small distortions is possible in principle, provided the complete GVFF is available. Whereas this is highly unlikely to be the case for compounds of unknown structure to which molecular mechanics is often applied, transferable diagonal elements should be aimed at, and to a first approximation all cross-terms ignored in the simulation. This is generally feasible for bond stretching and out-of-plane angle-bending modes since the harmonic approximation is rather poor for large torsions. The potential constants for torsional modes, however, are largely transferable as well.

The most common potential function for torsion in a bond simulates the electrostatic force between the charge distributions in the bonds adjacent to it. It was first suggested by Pauling[58], assuming partial *f*-character for an sp^3–sp^3 bond, that the function should be periodic and of 3-fold symmetry. It has been shown[59] that no *f* contribution is required to account for the barrier to rotation and hence that the 3-fold barrier in ethane relates to the molecular symmetry only. Potentials in common use like

$$U = \tfrac{1}{2} V(1 + \cos 3\Phi) \tag{26}$$

treat the torsion as a single interaction, assuming 3-fold symmetry as in ethane. In the absence of three-fold symmetry, the total interaction is better described in terms of nine individual contributions as shown in Fig. 2. However, these nine interactions are not simultaneously non-zero unless the potential is intrinsically periodic as suggested by Pauling[58]. It seems reasonable to assume that the torsional strain $U \to 0$ as $\Delta\Phi = |\phi - \phi_0| \to 60°$, and hence no more than three of the nine interactions in ethane is non-zero for any rotamer. This is correctly described by the function

$$U = \sum \frac{V}{6}(1 + \cos 3(\Delta\phi)) \tag{27}$$

$$\Delta < 60°\ .$$

In the case of delocalized interactions, the torsion as in a double bond is described by a potential function of the same form, but with $\phi_0 = 0°$, which implies an attractive interaction between adjacent bonds[68]. It follows that all $U \geq 0$.

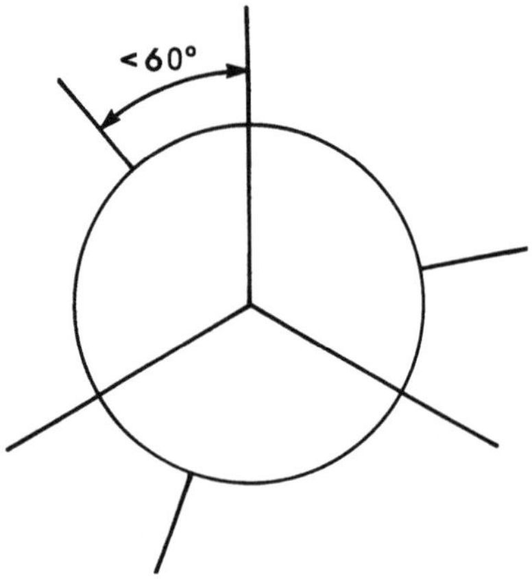

Fig. 2. Newman projection showing the nine possible torsional interactions in a bitetrahedral fragment. Only three of the torsion angles are less than 60° for any conformation

The interaction between atoms separated by more than two bonds are described in terms of potentials to represent non-bonded or Van der Waals interaction. A variety of potentials are used, but all of them correspond to attractive and repulsive components balanced to produce a minimum at an interatomic distance corresponding to the sum of the Van der Waals radii:

$$V_{nb} = R - A$$

The attractive component as a dispersive interaction between induced dipoles is formulated[61] as

$$A = c/r_{ij}^6$$

The repulsive component is less well defined, and the different formulae in use seem to depend on personal taste as much as anything. The most popular seem to be

(i) Lennard-Jones potential[62], $R = a/r_{ij}^{12}$
(ii) 9–6 potential[63], $R = a/r_{ij}^{9}$
(iii) Buckingham potential[64, 65], $R = a\exp(-br_{ij})$

The physical meaning of the parameters in these are compared by Abraham and Stölevik[66], who actually prefer a Morse function over all of them. The ready parametrization of the Buckingham function for a variety of atoms[67, 68] is no doubt responsible for its wide application. However, the fact that it becomes strongly attractive at short interatomic distances[69] as shown in Fig. 3 represents a distinct disadvantage in many applications.

Since it is impractical to restrict the use of molecular mechanics to compounds for which adequate force fields are available from vibrational analyses, approximate force fields based on totally different criteria inevitably come into being. Crystallographic results are one fertile source of force constants derived by trial-and-error methods aimed at a computational match of the observed molecular structure. Compared to the ideal of

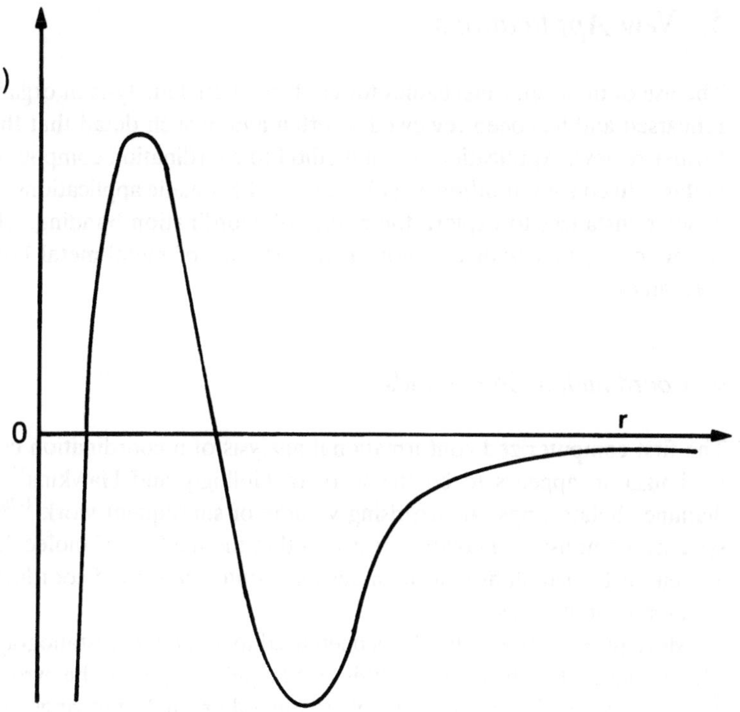

Fig. 3. Interatomic potential-energy curve according to a Buckingham potential, which fails at short interatomic distances

transferable force constants, this is a less desirable procedure since the results so obtained may be compound specific and of limited thermodynamic meaning. On the other hand, force constants calculated to reproduce measured thermodynamic quantities only may not be suitable for structure calculations.

To guard against this problem, Lifson and coworkers[70] introduced the idea of a consistent force field. A large set of experimental data – including vibrational spectra, structural properties, and thermodynamic measurements – is optimized by least-squares procedures in terms of force-field parameters for related families of compounds[71, 72].

This is about the present state of the art. New force fields are developed all the time with a variety of methods at about the same level of sophistication as outlined above. The method in its present form has reached its useful limits, and new developments are not anticipated until more realistic assumptions are adopted. In order to extend the method of molecular machanics to, for instance, the study of reaction pathways, the harmonic approximation, in particular, needs serious revision. Only a bonding curve like a Morse potential would allow the rupture and formation of bonds during computer simulation. For a general treatment of all bond types and compounds, theoretical methods for the quantum-mechanical estimation of strain-free (electronic) parameters are required.

The most useful approach for this purpose would not be the molecular-orbital method, but rather one that treats individual bonds in isolation, such as the bond-orbital methods discussed in the next section.

5. *New Applications*

The use of molecular mechanics for conformational analysis in organic chemistry is well rehearsed and has been reviewed so often and in such detail that there is little point in further review. Application of the method to coordination compounds is of more recent vintage. In concept it differs very little from the organic applications, but it has been used in some instances to explore the nature of coordination bonding. These aspects will be reviewed together with the more recent studies of metal-metal bonding by molecular mechanics.

a. Coordination Compounds

The first computerized conformational analysis of a coordination compound by energy minimization appears to be the work of Gollogly and Hawkins[73)] on five-membered diamine chelate rings. A surprising volume of subsequent work[74–78)] dealt with related systems, demonstrating quite effectively that the method of molecular mechanics works as well for the simulation of the molecular conformations of coordination compounds as for organic molecules.

More recently the method has been used to augment crystallographic studies in cases where standard diffraction analysis either fails or yields the wrong structure due to disorder effects. The moieties involved are either small, flat rings or disc-like, medium-sized cyclic species, as in the nickel perchlorate complexes of the fourteen-membered (N_4) macrocycles[79, 80)]. These latter structures are clearly disordered, and resolution of the disorder was possible only by calculating molecular mechanical trial structures that could be superimposed and refined crystallographically as separate rigid bodies. The disordered individuals, after refinement, were found to be related by local symmetry elements.

In other cases the symmetry due to disorder coincides with crystallographic symmetry elements[81)]. This allows resolution by the refinement of a single rigid body, representing an asymmetric unit of the disordered whole. If the molecular symmetry is sufficiently high, the disorder in these structures can effectively be disguised altogether. The molecules appear to be of higher symmetry, although the inevitable mismatch between superimposed symmetry-related individuals produces highly unlikely molecular geometries. The strain energies of these averaged arrangements considered as molecules are, likewise, also impossibly high. Such propositions nevertheless find their way into the literature as demonstrated[82)] in the case of the nickel perchlorate complex of diazacyclooctane. The crystallographically averaged, observed electron density is adequately accounted for in terms of the composite arrangement shown stereoscopically in Fig. 4, as calculated molecular mechanically.

Disorder of this type occurs because of two energetically equivalent, but geometrically different orientations of molecules in a crystal. The resulting mode of randomly disordered packing occurs without modification of molecular conformation.

The very common disordered arrangement of coordination rings in metallocene structures defines a closely related situation, but one more susceptible to intermolecular interactions. As a consequence, observed conformations are not reproduced by standard molecular mechanics, which applies to isolated molecules only. The observed disorder

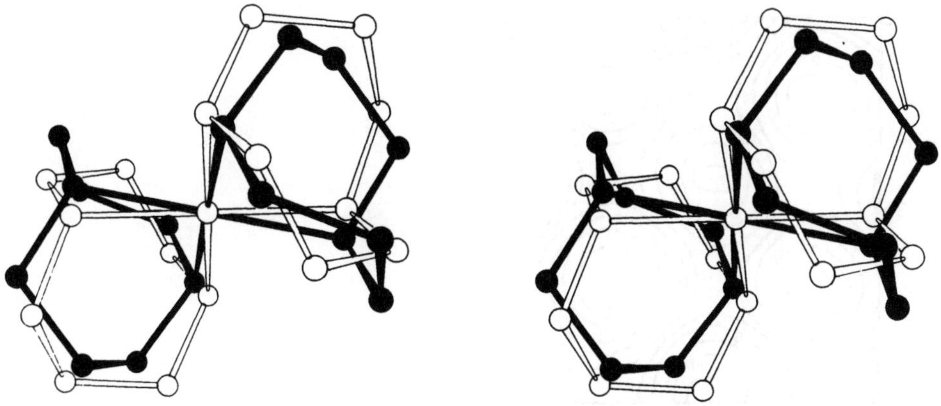

Fig. 4. Stereoscopic view of the molecular disorder in crystals of nickel perchlorate-diazacyclo-octane. These forms are not resolved in a standard crystallographic analysis

can be simulated by also taking into account crystal-packing forces in the form of intermolecular, non-bonded interactions. An obvious procedure is to calculate the total cohesive energy as a function of the orientation of the individual rings of a reference molecule embedded in the crystalline environment. This so-called fixed-lattice model was shown[83] not to discriminate correctly between minimum-energy arrangements. The reason for this is that the reorientation of molecules in a crystal is a cooperative effect and not allowed for individual entities in a fixed lattice.

An improved simulation must therefore allow concerted rotation of neighbouring rings like a set of mating gears. As an immediate consequence, the preservation of translational symmetry during ring rotation is implied. A model of this type was used successfully to simulate the disorder in the room-temperature ferrocene structure, to account for its phase transition on cooling, and to rationalize the different behaviour of nickelocene[84].

All of these effects are conveniently described in terms of a section through the potential-energy surface as a function of ring rotation, shown in Fig. 5. The rotating rings are of neighbouring sandwich units, and the rotations are measured in degrees away from the staggered configurations of both units, which appear to dominate the room-temperature structure.

The sub-minima at (0°, 36°) and (36°, 0°) correspond to mixed staggered and eclipsed configurations in the same structure. The sub-minimum at (36°, 36°) corresponds to the all-eclipsed structure. All four of the low-energy conformations are represented in the room-temperature crystal, but not in random disorder, since these forms are of low energy only in the concerted rotation model. This implies synergism between a molecule and its environment and disorder arising from a random distribution of ordered microdomains. The model predicts the correct, all-eclipsed low-temperature structure and a multistep phase transition, as observed experimentally[85]. Because of high-energy clashes during all combinations of concerted rotation, similar sub-minima do not occur for nickelocene, which hence has an ordered structure at room-temperature and no order-disorder phase transformation on cooling.

Ignoring the metal atoms shows that the packing in ferrocene closely relates to that of many other crystals containing flat molecules, like aromatic hydrocarbons and organic

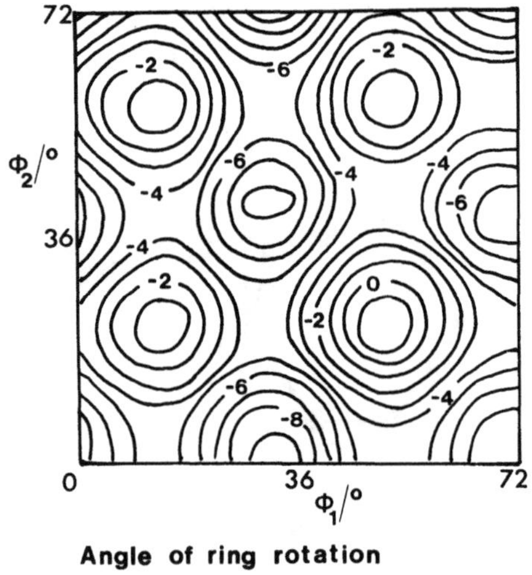

Fig. 5. Potential-energy environment as a function of concerted ring rotation in crystals of ferrocene

donor-acceptor or charge-transfer complexes. Related modes of disorder are therefore predicted and found in crystals of these compounds. A detailed molecular-mechanics analysis based on concerted rotation has been reported for the disorder in crystals of the anthracene-tetracyano-benzene (TCNB) charge-transfer complex[86]. Calculations predicted two energetically equivalent orientations of anthracene for a fixed orientation of TCNB. The corresponding disordered trial structure in which the contributing forms are related by a crystallographic mirror plane was successfully refined with previously published[87] diffraction data.

Another aspect of coordination chemistry studied in considerable detail by molecular mechanics is the macrocyclic effect. It readily explains the enhanced ligand-field strength and macrocyclic enthalpy in complexes of nitrogen-donor macrocycles[88] and provides an adequate model of ligand-hole sizes in these compounds[89, 90]. Apparent anomalies in metal-ligand bond lengths are shown not to relate to compression by the macrocycle, but rather to standard effects of steric interaction.

b. Dimetal Centres

All molecular-mechanical analyses of coordination compounds ignore non-bonded interactions involving metal atoms, presumably because the central metal atom is assumed to have an unimportant symmetrical dispersive effect, but more likely because appropriate interaction constants are difficult to generate. For most compounds with covalent dimetal centres, the same assumption would appear to be reasonable. However, a large number of non-bonded dimetal centres are known to occur in molecules like the dimeric carboxylates of copper and cadmium. To analyse these by force-field methods, it becomes essential to take non-bonded interactions involving metal atoms into account. The strategy used[60] to generate non-bonded potential functions for metal-atom interactions was to choose Van der Waals radii for transition-metal atoms to reproduce the known structures of dimeric carboxylates in molecular mechanics. It transpired that a variety of structures could be reproduced well with radii of 2.5 Å for first-row and 2.6 Å for other transition elements.

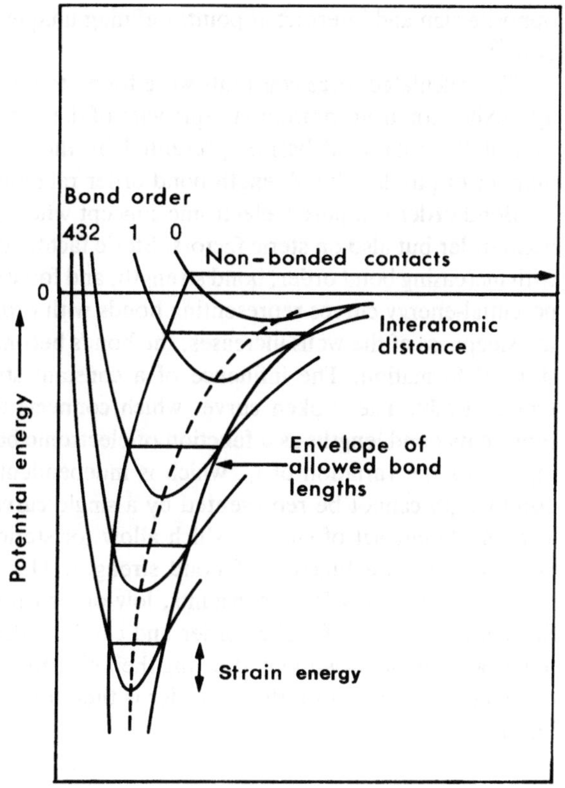

Fig. 6. Potential-energy curves of bonds differing in order only. The curve through the minima represents the variation of bond length as a function of bond order only. The effects of steric deformation are shown by the outer envelope

The results readily explain the stability of these systems and the unexpected proximity of the non-bonded metal atoms. It concerns the ready compressibility of metallic electron clouds compared to that of representative atoms. The Van der Waals cohesion in the highly symmetrical dicarboxylate framework, with a maximum at a characteristic bite, is sufficient to compact the non-bonded dimetal centre to an arrangement reminiscent of covalence. Most of the strain in the molecule accumulates between the metal atoms. Bulky ligands in the axial positions resist this compression to produce distortions that can easily be modelled by molecular mechanics.

Application of the method to dimetal bonds requires the specification of either k or r_0 and fixing of the other by trial-and-error matching of the observed bond length. In general this is not feasible. In axially substituted dichromium carboxylates, assignment is confused by the extreme variability of dimetal bond length. These bonds are consequently characterized[90)] by solution sets $\{k, r_0\}$. Although empirical force-constant relationships like Badger's rule[91)] proved useful in sampling the solution sets, this is not an essential procedure since the non-crossing solution curves clearly define the important relative trends.

An interesting situation exists for the $M \overset{4}{-} M$ bonds of Cr, Mo and W. Apart from the dicarboxylates, where the metal centres occur in sterically cohesive environment, they are also present as sterically stretched bonds in the octamethyl dimetal anions, $M_2(CH_3)_8^{4-}$. The solution curves (k vs r_0) of the two types of compound have slopes of opposite sign and intersect at points defining unique solutions (k, r_0) for each quadruple bond[92].

The calculated force constants were found to match the values obtained by harmonic approximation from vibration frequencies of the unbridged species. This enabled calculation of the rotational barrier presented by the δ-component and served to clarify a number of puzzling bond length-bond order relationships identified before[93].

Bond order is a purely electronic concept whereas bond length depends not only on bond order but also on steric factors. Steric factors become progressively less important with increasing bond order, bond strength, and force constant. Figure 6 shows a family of potential-energy curves representing bonds with order increasing from zero to four. As the steepness of the wells increases, the bonds become less flexible and more resistant to steric deformation. The influence of a constant strain on various bonds is indicated schematically. The broken curve, which connects the minima of the bonding curves, represents bond lengths as a function of electronic bond order only and hence it actually represents the variation of r_0, which is independent of steric factors. The variation of bond length cannot be represented by a single curve because of its steric dependence. The enveloping set of curves, which allow for steric effects, defines a field of possible bond lengths as a function of bond strength. The most important conclusion is that, because of their low force constants, low-order (including zero order) bonds are more flexible than bonds of higher order and can be either compressed or stretched by relatively large amounts without affecting their electronic bond order. An overlap of accessible lengths of bonds of different order is therefore predicted as a consequence of steric interactions.

IV. Bonding Theories

In molecular mechanics the theory of chemical bonding is relied on to provide the strain-free parameters and constants of the force field, although this exercise is rarely carried through. In current applications, harmonic potential-energy curves are assumed for all chemical bonds and angles despite the fact that in numerous cases distortions far in excess of anything allowed by the harmonic model are known to occur[94]. A need therefore exists for theoretical models to provide potential-energy curves for chemical bonds in molecules. The theory should concentrate primarily on the bonding region and the nature of the atoms directly involved. In principle this is a feasible approach[95]. Not only are there many examples of bonding properties calculated by valence-electron-only methods[96], it is also well established that electrons in molecules separate more-or-less neatly into independent lone-pair, bonding and core loges[97]. In essence the required theory therefore reduces to the traditional electron-pair view of chemical bonding.

In the language of quantum logic, this model amounts to an abstraction of bonding electrons, one per atom, in the field of classical cores. This ignores EPR correlation between bonding electrons and cores, but not electromagnetic interaction. This means that two bonded atoms in a molecule can be viewed as monopositive cores interacting quantum-mechanically with the bonding pair of electrons and classically with each other, that is, electrostatically. The behaviour of the quantum-mechanical bonding electron is described by an appropriate wavefunction which is not an atomic wavefunction and which can conveniently be called a bond orbital.

1. *Bond-Orbital Methods*

In bond-orbital formalism as outlined above the bonding problem reduces to the level of the diatomic molecule with the bond orbital as an unknown. Density-functional theory suggests a possible solution. According to the Hohenberg-Kohn theorem[99], a given ground-state electron density determines its ground-state wavefunction and consequently the properties of the system, even though there are an infinite number of anti-symmetric functions that give the density. Practical calculations within density-functional theory usually appeal to the Kohn-Sham theorem[100], which defines the density and energy in terms of an appropriate self-consistent Schrödinger equation. Perdew and others[101] showed that exact results may be derived for the maximum occupied Kohn-Sham orbital energy, which is identified as the exact ionization energy when density-functional theory is extended to fractional electron number, i.e.

$$-I_p = \varepsilon_{max} = -\frac{x}{n^2}, \text{ for } 0 < f < 1 \ ,$$

where n is the principal quantum number. For an atom of given radius, σ, this becomes[102]

$$I_p = \frac{k}{n^2\sigma}$$

This also describes the electron that features in the formation of electron-pair bonds. The density relates its wavefunction to the first atomic ionization potential, and this should be reflected in the definition of bond orbitals. In the absence of any further information, the most reasonable procedure is to assume uniform bonding-electron density ϱ over a spherical volume of radius σ and to define a wavefunction $\psi^2(r) = \varrho$ and hence

$$\psi(r) = \left(\frac{f}{V}\right)^{1/2} = (3f/4\pi\sigma^3)^{1/2}; 0 < r < \sigma \ . \quad (30)$$

The assumed functional dependence implies that the ionization potential can be formulated in terms of electron number and density,

$$I_p = k'' \cdot \varrho \cdot f = \frac{k'f^2}{\sigma^3} \ .$$

Hence $\frac{nf}{\sigma} = c = \sqrt{k/k'}$.

It follows that $\psi(r) = (3\,c/4\,\pi n\sigma^2)^{1/2}$; $0 < r < \sigma$, or in general

$$\psi(r) = (3\,c/4\,\pi n)^{1/2}(1/\sigma)\exp[-(r/\sigma)^p] \tag{31}$$

where $p \gg 1$.

This wavefunction was first obtained empirically[183] with atomic radii derived from Hartree-Fock atomic wavefunctions and without appreciating the relationship with ionization potential. Although deliberate efforts to use this model for the derivation of bond parameters in molecular mechanics have not been made, sufficient preliminary results have been obtained to demonstrate the power and potential of the method. It reproduces bond lengths and energies[104], bond angles[105], the properties of multiple bonds[106] and force constants[107] with good accuracy. The parameter c is nearly constant for all atoms and accounts for the fact that bond properties are largely insensitive to environment and have transferable values, so useful in molecular mechanics.

2. *Electrostatic Methods*

An even simpler description of chemical bonding is possible in terms of electrostatic concepts. The correct approach is dictated by the theorem of Hellmann[41] and Feynman[42]. This theorem is only valid within the Born-Oppenheimer approximation, which separates nuclear and electronic motion. The nuclei can be considered as fixed and with zero kinetic energy contribution to the molecular Hamiltonian. The electronic-energy eigenvalue in the field of the fixed nuclei

$$E' = \int \psi^* H_e \psi dr$$

consitutes an effective potential for nuclear motion, and the force on a nucleus is calculated as

$$F_a = -\frac{\partial E'}{\partial r_\alpha} . \tag{32}$$

The x-component of the force follows as

$$F_{\alpha x} = -\int \psi^* \frac{\partial H_e}{\partial x_\alpha} \psi dr - \int \frac{\partial \psi^*}{\partial x_\alpha} H_e \psi dr - \int \psi^* H_e \frac{\partial \psi}{\partial x_\alpha} dr . \tag{33}$$

Since H_e is a real Hermitian operator

$$\int \psi^* H_e \frac{\partial \psi}{\partial x_\alpha} dr = \int \frac{\partial \psi}{\partial x_\alpha} H_e \psi^* dr$$

and since $H_e\psi^* = E'\psi^*$, the last two terms above can be rewritten as

$$-E'\left[\int \frac{\partial \psi^*}{\partial x_\alpha} \psi dr + \int \psi^* \frac{\partial \psi}{\partial x_\alpha} dr\right] = -E' \frac{\partial}{\partial x_\alpha} [\int \psi^* \psi dr] = 0 .$$

The kinetic energy of H_e does not depend on the nuclear coordinate x_α, and the force component becomes

$$F_{\alpha x} = -\int \psi^* \frac{\partial H_e}{\partial x_\alpha} \psi dr = -\int \psi^* \frac{\partial V}{\partial x_\alpha} \psi dr \tag{34}$$

where V is the classical potential for a set of nuclear and electronic coordinates as defined in (4). The forces on the nuclei can thus be calculated from the classical potential and the electronic wavefunction.

This result is in line with the definition of molecular structure and chemical bonding as classical concepts. From the electronic charge distribution, the bonding forces holding the nuclei together can in principle be calculated by the methods of classical electrostatics. This could be anticipated since the abstraction that separates electrons and nuclei requires a classical description of the molecular structure.

Since the resultant force on an atom in a molecule vanishes, this treatment will be restricted to equilibrium situations[108)]. Because of different possible modes of polarization, more than one equilibrium charge distribution can, however, occur for a given pair of atoms. This is usually described in terms of different orders of bonding. It means that for each conceivable bond order there is a characteristic equilibrium separation with a matching charge distribution. These equilibrium arrangements therefore occur along a curve such as the broken line in Fig. 6, which represents a continuously varying bond order.

On the basis of the previous assumption of uniform charge densities and characteristic atomic radii, any equilibrium situation is correctly described in terms of the point-charge model[109)] illustrated schematically in Fig. 7. The magnitude of the point charges depends on the volume of overlap and hence is a function of the equilibrium separation. This yields an interaction potential energy

$$E = K\left[\delta\varepsilon\left(\frac{1}{b} + \frac{1}{d+b} - \frac{1}{d} - \frac{1}{p}\right) + X\right] \tag{35}$$

K is a dimensional constant

$X = 0$, unless $r_2 > r_1 + d$, where $X = [(1 - r_1)/(r_2 - d)]^2/d$

$$\delta\varepsilon = \left[\frac{d^3}{16} - \frac{3d}{8}(r_1^2 + r_2^2) - \frac{3}{16d}(r_2^2 - r_1^2)^2 + \tfrac{1}{2}(r_1^3 + r_2^3)\right]^2/(r_1 r_2)^3 \tag{36}$$

$p = r_1 + r_2 - d$

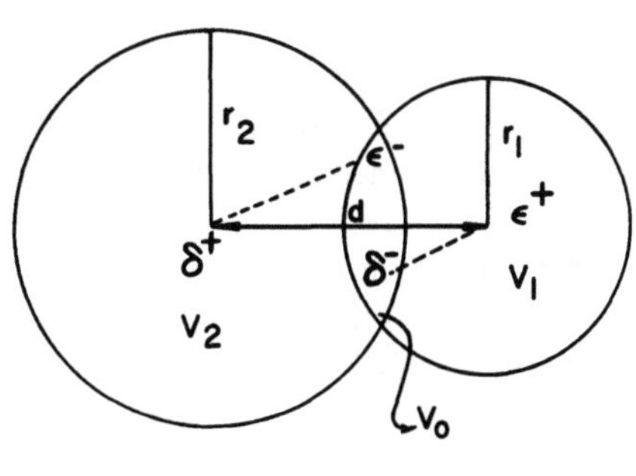

Fig. 7. Point-charge model representing a covalent bond in terms of overlapping electron clouds of uniform charge density

$b = \frac{1}{2}(r_1/r_2)d$, for $r_1 < r_2$

This model has been tested exhaustively[109, 110] for a large variety of bonds and found to provide a good description of bond strength as a function of bond length. The radii are the same as those used in the electron-pair model described above. It was actually demonstrated[109] that a linear relationship with ionization potential existed.

The energy expression for homonuclear bonds ($r_1 = r_2$) in atomic units (d in units of r and E in units of 1/r) reduces to

$$E = \frac{d^5}{64} + \frac{d^4}{128} - \frac{23\,d^3}{64} + \frac{11\,d^2}{32} + \frac{25\,d}{16} - 5 + 3\,d^{-1} \quad (37)$$

which graphically has the same form as the broken curve of Fig. 6. This confirms that the bond lengths in the electrostatic model correspond to the strain-free bond lengths of molecular mechanics. The relationship between bonds of different order is seen to be a simple screening function as discovered empirically before[106, 107].

An essentially equivalent procedure was used by Ohwada[111] to calculate valence force constants in polyatomic molecules. Instead of characteristic radii, however, effective nuclear charges were specified. The force constant for a heteronuclear diatomic molecule is first obtained as

$$k_{ij} = 2[Z_i e(1 - f_{ii})^{1/2}][Z_j e(1 - f_{jj})^{1/2}]/R_{ij}^3 = 2\,Z_i^* Z_j^*/R_{ij}^3 \quad (38)$$

where f represents a screening factor and R an equilibrium interatomic distance. With a diatomics-in-molecules approach, general expressions for all force and interaction constants for three-atom fragments were obtained in terms of interatomic distances and angles only. Agreement with experiment is satisfactory.

In principle a complete molecular-mechanics force field can thus be derived from simple bonding models although it remains to be demonstrated in practice.

V. Complications

Chemical theory requires the definition of molecular structure as a classical entity and a quantum-mechanical description of electron density. This is feasible only by suitable abstraction that ignores EPR correlation between electrons and nuclei. In this instance the sum of the parts is clearly less than the total, and the resulting description of the chemical system is perforce incomplete. The most serious deviation from predicted behaviour was first identified by Jahn and Teller[112]. They showed that if the electronic state of a symmetrical nuclear configuration is degenerate, then this configuration is unstable with respect to nuclear displacements that remove the degeneracy. If the molecular Hamiltonian is given by either (5) or (6), depending on whether or not nuclear kinetic energy is taken into account, the latter may be considered as a small perturbation. Since the perturbation approach is qualitatively different for the degenerate and non-degenerate cases, non-zero nuclear displacements are predicted for electronically degenerate systems only. This is the origin of Jahn-Teller and pseudo Jahn-Teller effects and of vibronic coupling effects in general.

1. Vibronic Interactions

The adiabatic approximation as a zero-order perturbation is discussed by Davydov[113)] and by Bersuker and Polinger[114)]. The Hamiltonian that determines the internal states of a system is written as

$$H = T(Q) + T(q) + V(q, Q) \tag{39}$$

where

$$T(q) = -\frac{h^2}{8\pi^2 m}\sum_i \frac{\partial^2}{\partial q_i^2} \tag{39}$$

$$T(Q) = -\frac{h^2}{8\pi^2 m}\sum_\alpha \frac{\partial^2}{\partial Q_\alpha^2}$$

and $V(q, Q)$ is the potential-energy operator between all the particles. $T(Q)$ is treated as a perturbation, whereby

$$H = H_0 + T(Q) .$$

As $M \to \infty$, the Schrödinger equation becomes

$$[H_0 - \varepsilon_n(Q)]\psi_n(q, Q) = 0 \tag{40}$$

for infinetely slow (adiabatic) changes in Q, the nuclear coordinates.

Stationary states using the total Hamiltonian

$$(H - E)\Psi(q, Q) = 0 \tag{41}$$

can be written in terms of the adiabatic eigenfunctions ψ_n, as

$$\Psi(q, Q) = \sum_n \Phi_n(Q)\psi_n(q, Q) . \tag{42}$$

Substituting (42) into (41), multiplying by ψ_m^*, and integrating over the electronic coordinates, q, gives

$$[T(Q) + \varepsilon_m(Q) - E]\Phi_m(Q) = \sum_n \Lambda_{nm}\Phi_n(Q) \tag{43}$$

where the non-adiabacity operators are

$$\Lambda_{mn} = \frac{h^2}{4M^2}\sum_\alpha \int \psi_m^*(q, Q)\frac{\partial}{\partial Q_\alpha}\psi_n(q, Q)dq\frac{\partial}{\partial Q_\alpha} - \int \psi_m(q, Q)T(Q)\psi_n(q, Q)dq .$$

In the adiabatic approximation the operator Λ_{mn} is assumed to be zero, giving

$$[T(Q) + \varepsilon_m(Q)]\Phi^0_{mv}(Q) = E^0_{mv}\Phi^0_{mv} \tag{44}$$

where v are nuclear quantum numbers, and the wavefunction of the system reduces to

$$\Psi_{mv} = \Phi^0_{mv}(Q)\psi_m(q, Q) . \tag{45}$$

Perturbation theory shows that the adiabatic approximation is valid only for

$$\int \Phi^0_{mv}\Lambda_{mn}\Phi^0_{mv'} \ll |E^0_{mv} - E^0_{nv'}| . \tag{46}$$

An estimate in terms of the largest nuclear vibration frequency ω, near the potential-energy point, yields the criterion

$$\frac{h\omega}{2} \ll |E_m - E_n| = \Delta E_{el} \tag{47}$$

In symmetric molecular systems containing rotation axes $C_n\,(n \geq 3)$, have two or several different electron distributions with the same energy (electronic degeneracy) for which $\Delta E = 0$. This is the Jahn-Teller effect.

It means that the adiabatic approximation breaks down for symmetrical nuclear arrangements that produce electronic degeneracy. The non-adiabatic response removes the degeneracy and distorts the symmetrical nuclear arrangement. It is not only degenerate terms, but also quasi- or pseudo-degenerate energy terms that cannot be treated by the adiabatic approximation. Many of these related effects are often also referred to as Jahn-Teller effects. It is emphasized by Bersuker and Polinger[114)] that, in general, deviations from the adiabatic approximation are due to the mixing of different electronic states by nuclear displacements (by vibronic interaction terms in the Hamiltonian), and this vibronic mixing is stronger the closer in energy the mixed states are and the greater the appropriate vibronic constant.

The real meaning of vibronic effects concerns the behaviour of different adiabatic potential surfaces near the point of their intersection, where they have no physical meaning, due to the breakdown of the adiabatic approximation. Although the consequences of vibronic coupling are considered to be largely outside the scope of this review, they can evidently not be ignored in the future development of molecular mechanics. This method in its present form inevitably reproduces the minimum-energy arrangements at which vibronic effects become important. However, until convenient methods become available for the solution of the vibronic equations and the calculation of adiabatic potential surfaces with sufficient accuracy to allow prediction of nuclear displacement, these effects cannot conveniently be taken into account.

2. Deformation Densities

It almost goes without saying that the effect of vibronic coupling on electron density should be more severe than on the positions of the more massive nuclei. Non-adiabatic responses of this type could obviously invalidate the model used for the interpretation of deformation densities commonly assumed to represent chemical bonding effects only. The basic assumption of the model is the minimal distortion of atomic core-electron density at molecule formation[115, 116]. This is formulated in terms of X-ray scattering as[117]

$$\Delta F(\text{core}) = F(\text{core, molecule})\text{-}F(\text{core, promolecule}) \simeq 0\ .$$

By separating the total scattering into core and valence contributions, one has

$$F(\text{total}) = F(\text{core}) + F(\text{valence})$$

and hence that

$$\Delta F(\text{total}) = \Delta F(\text{valence})\ .$$

On this basis the molecular charge density is written[108] as

$$\varrho(\text{mol}) = \varrho(\text{promol}) + \delta\varrho$$

$$\text{where}\quad \varrho(\text{promol}) = \sum_{\alpha} \varrho_{\alpha}^{\text{at}}$$

is the sum of spherically symmetric, free-atom densities centred at the molecular atomic positions. The deformation density $\delta\varrho$ is interpreted as the valence charge migration whereby the collection of non-interacting atoms is converted into a chemically bound molecule.

This is essentially the same assumption made in the bond-orbital method already described: hence it relies on abstracting valence electrons in the field of classical cores. The abstraction ignores EPR correlation, vibronic coupling, and singularities in the Born-Oppenheimer surface. An inherent danger is that these factors could lead to significant distortions of the electronic cores. Arguments to discount the importance of core deformation are largely qualitative[108]. Since the total deformation density, however, is the small difference between large quantities, slight core distortions could easily be interpreted as major chemical effects. This abstraction could be at the root of the fundamentally different interpretations of the role of electron density in chemical bonding presented by Dunitz and Seiler[118] as opposed to the presentation of Berlin[119].

Dunitz and Seiler[118] studied the deformation density in a centrosymmetric, polycyclic molecule of composition $C_6H_{12}N_4O_4$ by standard X-ray electron-density difference mapping. They found a steady decrease in the deformation densities within bonds, in the order C–N > C–O > N–N > O–O, and of such magnitude that the density along the O–O bond is negative throughout. A general charge deficit in bond regions between electron-rich atoms is inferred and attributed to electron dispersal due to the exclusion principle. It is argued that with the sole exception of the atypical H_2 molecule, an accumulation of charge between atoms is anti-bonding.

This is in stark contrast with the findings of Berlin[119)] who calculated boundary surfaces between electrostatic binding and antibinding zones between atoms in molecules on the basis of the Hellman-Feynmann theorem. For homonuclear pairs, the boundary curves are hyperbolae with straight-line asymptotes intersecting with the tetrahedral angle, 109°28′. This is practically the inverse of the difference density map obtained by Dunitz and Seiler[118)] for the O–O region.

In terms of simple bonding theory it appears very reasonable that the deformation density could be negative in some bonds. When a p-block atom with n valence electrons forms a bond, the valence shell is polarized into a tetrahedral distribution with n/4 electrons at each potential bonding site. Since only one electron is required to form each bond, the predicted deformation density of 2(1 − n/4) electrons per bond is exactly in line with the observed[118)] trends. However, polarization only occurs when a bond forms, and the free-atom density should therefore be averaged, not over four potential bonding sites, but over the total solid angle, i.e.

$$\bar{\varrho} = \frac{n}{4\Pi} \text{ e/sterad} .$$

Furthermore, polarization produces, not point charges, but as shown in Fig. 8, a spread of bonding electron density over one of four contiguous solid angles of $3\pi/4$ sterad each, and the density varies with linear angle:

$$\varrho = \varrho_0 f(\theta) .$$

The total charge within a solid angle of radius a is

$$\varrho_0 \int_{-a}^{a} f(\theta) d\theta = 1 .$$

For sp^3 orbitals the density varies almost like $\cos^2\theta$, and for $a = 3\pi/8$ the maximum density at $\theta = 0$ follows as

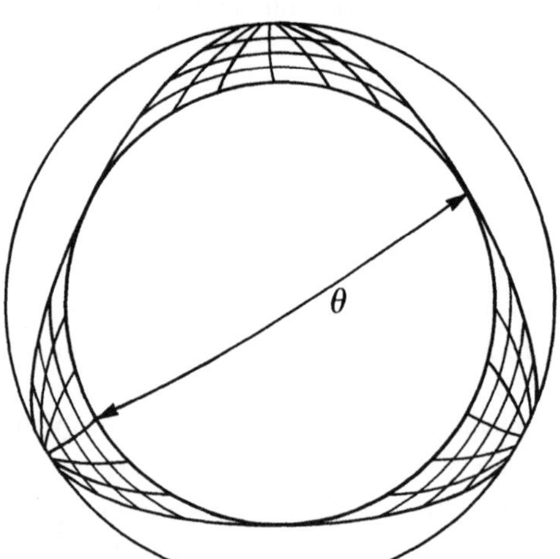

Fig. 8. Four contiguous conical solid angles shown at the surface of a sphere. Excluded volumes are hatched

$$\varrho_0 = \frac{1}{\left[\frac{3\pi}{8} + \frac{1}{2}\sin\left(\frac{3\pi}{4}\right)\right]} = 0.6529 \text{ e/sterad} .$$

Since $\varrho_0 > \bar{\varrho}$ for all $n < 9$, positive deformation densities are predicted in all bonds of interest. ($\bar{\varrho}_9 = 0.7162$.)

The discrepancy is related to the work of Ermler and Kern[120)] who calculated electron densities in the water molecule at 45 different geometries along the normal-mode coordinates in order to assess the effect of zero-point vibrations on the density. They actually found poor correlation between the density shifts at the nucleus and for the entire atom. For oxygen, surprisingly, they also found a density increase at the nucleus, compared to the atom at rest. Extrapolation of these results to thermal vibrations suggests how the spherical free-atom scattering model becomes increasingly inadequate at non-zero temperatures. The deformation density in the region between two oxygen atoms would therefore be particularly severely affected by these distortions due to vibronic coupling.

More recently Hirshfeld[121)] showed that a discrepancy, not supported by theoretical calculations on related compounds, is observed in the deformation mapping of the C–F bond in tetrafluoroterephthalonitrile. The conclusion that the stability of the C–F bond is understood theoretically, but completely unverifiable by X-ray diffraction because of an undetectable forward polarization of the core region around F, is in line with the explanation in terms of vibronic coupling.

It is tempting to conclude that in view of the approximate nature of the abstraction required to isolate valence electrons, distortion of atomic electron cores and non-adiabatic EPR correlations can no longer be ignored in accurate electron-density studies, since anomalies such as negative bonding densities due to the neglect of vibronic coupling are in fact predicted.

VI. Conclusion

The chemist thinks in terms of molecules and, from diffraction studies, gets to know their geometrical shapes. Nuclear positions are even directly observable in neutron diffraction. However, this interpretation of diffraction patterns has meaning only in terms of an assumed quasi-static arrangement of atomic nuclei. More abstract interpretations in terms of symmetry operators only are not rigorously excluded. Molecular structures are observed only in order to simplify the interpretation of chemistry and not because they exist. The detection of molecular structures becomes feasible only in terms of appropriate abstractions, which imply inevitable approximations.

Use of the observed structures to synthesize chemical behaviour magnifies the defects of the abstraction. Anomalous features like Jahn-Teller distortions or negative bonding-electron densities are some of the consequences. More seriously the only permissible mathematical model in abstraction is necessarily restrictive because it requires a commutative algebra of observables. Wherever the molecular-structure hypothesis is invoked, the enquiry becomes restricted to classical concepts, and inability to demonstrate quantum effects should not come as a surprise. This is an error of abstraction that is introduced by the analytical procedure, as outlined by Jan Smuts[122)]:

"Analysis, abstraction and generalisation are indeed necessary as instruments of scientific understanding, but they also necessarily involve a departure from the complex concrete, and thus produce a possible element of error which in its ultimate effects may produce a serious distortion in our general view of reality. The concrete whole of a situation comes to be deduced from its abstracts, and the principle of natural explanations proceeds by way of the parts to the whole. The whole as so understood is confined to its parts and comes to suffer from the same limitations as its parts. For the full concrete reality comes to be substituted a more limited scheme or pattern of parts, an aggregation rather than a natural organic synthesis."

Molecular mechanics is the vehicle of the structure hypothesis. It provides a coherent framework for describing the classical attributes of molecules, but it cannot account for the holistic behaviour of chemical systems, which revokes the molecular-structure hypothesis.

Acknowledgements. This has been a sabbatical project, and I gratefully acknowledge financial support from:
The Council for Scientific and Industrial Research;
The Ernest Oppenheimer Memorial Trust;
The Anderson Capelli-Convocation Fund; and
The University of the Witwatersrand for granting me a year's leave.

Many of the ideas first took shape while working in the stimulating environment of the Laboratory for Molecular Structure and Bonding under the direction of F. Albert Cotton at Texas A & M University, College Station, Texas. I acknowledge extremely useful discussions with John P. Fackler, Larry Falvello and Scott Han at A & M and with Tony Ford and Demetrius Levendis at Wits.

I thank Professor John B. Goodenough for many helpful suggestions.

VII. References

1. Casimir, H. G. B.: Haphazard Reality, Harper and Row, New York, 258 (1983)
2. Gillespie, R. J.: J. Chem. Ed. *47,* 18 (1970)
3. Claverie, P.: in: Symmetries and Properties of Non-rigid Molecules (eds. Maruani, J., Serre, J.), Elsevier, Amsterdam 1983
4. Avogadro, A.: J. Phys. Paris *73,* 58 (1811)
5. Kekule, F. A.: Z. f. Chem. *3,* 643 (1860)
6. Butlerov, A. M.: Bull. Soc. Chim. Fr. *5,* 582 (1863)
7. Van't Hoff, J. H.: La Chimie dans L'Espace, Barzendijk, Rotterdam 1875
8. Pasteur, L.: Ann. Chim. Phys. *24,* 442 (1848)
9. Le Bel, J. A.: Bull. Soc. Chim. *22,* 237 (1874)
10. Prelog, V.: Science *193,* 17 (1976)
11. Bohr, N.: Phil. Mag. *26,* 1, 476, 857 (1913)
12. Woolley, R. G.: Adv. Phys. *25,* 27 (1976)
13. Schwarzschild, K.: Sitzungsber. preuss. Akad. Wiss. *1,* 548 (1916)
14. Heurlinger, T.: Z. Phys. *20,* 188 (1919)
15. Franck, J.: Trans. Faraday Soc. *21,* 536 (1925)
16. Condon, E. U.: Phys. Rev. *27,* 640 (1926)
17. Mulliken, R. S.: Phys. Rev. *26,* 561 (1925)
18. Mulliken, R. S.: Proc. Nat. Acad. Sci, U.S.A. *12,* 144, 151, 158, 338 (1926)
19. Heitler, W., London, F.: Z. Phys. *44,* 455 (1927)
20. Wilson, E. B.: Int. J. Quantum Chem. Symp. *11,* 17 (1977)

21. Coulson, C. A.: Valence, 2nd ed., Oxford University Press 1961
22. Born, M., Oppenheimer, R.: Ann. Phys. *84,* 457 (1927)
23. Born, M., Huang, K.: Dynamical Theory of Crystal Lattices (eds. Born, M., Huang, K.), Clarendon, Oxford 1954
24. Slanina, Z.: Adv. Quantum Chem. *13,* 89 (1981)
25. Woolley, R. G.: Struct. and Bonding *52,* 1 (1982)
26. Woolley, R. G.: Chem. Phys. Lett. *55,* 443 (1978)
27. Claverie, P., Diner, S.: Isr. J. Chem. *19,* 54 (1980)
28. Hagedorn, G. A.: Commun. Math. Phys. *77,* 1 (1980)
29. Berry, R. S.: in: Ref. 32, p. 143
30. Dennison, D. M., Uhlenbeck, G. E.: Phys. Rev. *41,* 313 (1932)
31. Kilpatrick, J. E., Pitzer, K. S., Spitzer, R.: J. Am. Chem. Soc. *69,* 2483 (1947)
32. Woolley, R. G. (Ed.): Quantum Dynamics of Molecules: The New Experimental Challenge to Theorists, Nato ASI, Series B: Physics, Vol. 57. Editor's Preface 1979
33. Primas, H.: in: Ref. 32, p. 39
34. Primas, H.: Chemistry, quantum mechanics and reductionism – perspectives in theoretical chemistry, Lectures notes in Chemistry, Springer, Berlin, 1981, *24*
35. Einstein, A., Podolsky, B., Rosen, N.: Phys. Rev. *47,* 777 (1935)
36. Bell, J. S.: Physics *1,* 195 (1965); Rev. Mod. Phys. *38,* 447 (1966)
37. Clauser, J. F., Shimony, A.: Rep. Prog. Phys. *41,* 1881 (1978)
38. D'Espagnat, B.: Phys. Rev. *D11,* 1424 (1975); D'Espagnat, B.: Conceptual Foundations of Quantum Mechanics, Benjamin, Reading, Mass., 2nd ed., 1976
39. D'Espagnat, B.: Sci. Am. 128, Nov 1979
40. Raggio, G. A.: States and composite systems in W* algebraic quantum mechanics, Dissertation ETH No. 6824, Zürich 1981
41. Hellmann, H.: Einführung in die Quantenchemie, Deuticke, Leipzig 1937
42. Feynman, R. P.: Phys. Rev. *56,* 340 (1939)
43. White, D. N. J.: Mol. Struct. Diffr. Methods, *6,* 38 (1978)
44. Dośen-Mićović, L., Jeremić, D., Allinger, H. L.: J. Am. Chem. Soc. *105,* 1716 (1983)
45. Morse, P. M.: Phys. Rev. *34,* 57 (1929)
46. Wilson, E. B., Decius, J. C., Cross, P. C.: Molecular Vibrations, McGraw-Hill, New York 1955
47. Woodward, L. A.: Introduction to the theory of molecular vibrations and vibrational spectroscopy, Clarendon, Oxford 1972
48. Heicklen, J.: J. Chem. Phys. *36,* 721 (1962)
49. Aldous, J., Mills, I. M.: Spectrochim. Acta *18,* 1073 (1962)
50. Schachtsneider, H., Snyder, R. G.: Spectrochim. Acta *19,* 117 (1963)
51. Aron, J., Bunnell, J., Ford, T. A., Mercau, N., Aroca, R., Robinson, E. A.: J. Mol. Struct. *110,* 361 (1984)
52. Zerbi, G.: in: Vibrational Spectroscopy – Modern Trends (eds. Barnes, A. J., Orville-Thomas, W. J.), Elsevier, Amsterdam, 261 (1977)
53. Cotton, F. A., Kraihanzel, C. S.: J. Am. Chem. Soc. *84,* 4432 (1962)
54. Williams, J. E., Stang, P. J., von Schleyer, P.: Annu. Rev. Phys. Chem. *19,* 531 (1968)
55. Dennison, D. M.: Phil. Mag. *1,* 195 (1926)
56. This approach is used in the crystallographic refinement program SHELX – Sheldrick, G. M.: in: Computing in Crystallography (eds. Schenk, H., Olthof-Hazekamp, R., Van Koningsveld, H., Bassi, G. C.), Delft University Press 1978
57. Shimanouchi, T.: IUPAC Special Lectures, Molecular Structure and Spectroscopy, Butterworths, 131 (1963)
58. Pauling, L.: Proc. Natl. Acad. Sci. U.S.A. *44,* 211 (1958)
59. Pitzer, R. M.: Acc. Chem. Res. *16,* 207 (1983)
60. Boeyens, J. C. A., Cotton, F. A., Han, S.: Inorg. Chem. *24,* 1750 (1985)
61. Pitzer, K. S.: Adv. Chem. Phys. *2,* 59 (1959)
62. Lennard-Jones, J. E.: Physica *4,* 941 (1937)
63. Warshel, A., Lifson, S.: J. Chem. Phys. *53,* 582 (1970)
64. Born, M., Mayer, H. E.: Z. Phys. *75,* 1 (1932)
65. Buckingham, R. A.: Proc. Roy. Soc. (Lond.) *A168,* 264 (1938)

66. Abraham, J., Stölevik, R.: Chem. Phys. Lett. *58,* 622 (1978)
67. Brant, D. A., Flory, P. J.: J. Am. Chem. Soc. *87,* 2788 (1965)
68. Scott, R. A., Scheraga, H. A.: J. Chem. Phys. *42,* 2209 (1965)
69. Hirshfelder, J. O., Curtiss, C. F., Bird, R. B.: Molecular Theory of Gases and Liquids, Wiley, New York 1954
70. Lifson, S., Warshel, A.: J. Chem. Phys. *49,* 5116 (1968)
71. Ermer, O., Lifson, S.: J. Am. Chem. Soc. *95,* 4121 (1973)
72. Ermer, O.: Struct. and Bonding *27,* 161 (1976)
73. Gollogly, J. R., Hawkins, C. J.: Inorg. Chem. *8,* 1168 (1969)
74. Snow, M. R.: J. Am. Chem. Soc. *92,* 3610 (1970)
75. Niketic, S. R., Rasmussen, K., Woldbye, F., Lifson, S.: Acta Chem. Scand. *A30,* 485 (1976)
76. Yoshikawa, Y.: Bull. Chem. Soc. J. *49,* 159 (1976)
77. Geue, R., Snow, M. R.: Inorg. Chem. *16,* 231 (1977)
78. McDougall, G. J., Hancock, R. D., Boeyens, J. C. A.: J. Chem. Soc. Dalton Trans., 1438 (1978)
79. Boeyens, J. C. A.: Acta Crystallogr. *C39,* 846 (1983)
80. Thöm, V. J., Fox, C. C., Boeyens, J. C. A., Hancock, R. D.: J. Am. Chem. Soc. *106,* 5947 (1984)
81. Boeyens, J. C. A., Hancock, R. D., Thöm, V. J.: J. Cryst. Spectr. Res. *14,* 261 (1984)
82. Boeyens, J. C. A., Fox, C. C., Hancock, R. D.: Inorg. Chim. Acta *87,* 1 (1984)
83. Levendis, D. C., Boeyens, J. C. A.: S. Afr. J. Chem. *35,* 144 (1982)
84. Levendis, D. C., Boeyens, J. C. A.: J. Cryst. Spectr. Res. *15,* 1 (1985)
85. Edwards, J. W., Kington, G. L., Mason, R.: Trans. Faraday Soc. *56,* 660 (1960)
86. Boeyens, J. C. A., Levendis, D. C.: J. Chem. Phys. *80,* 2681 (1984)
87. Tsuchiya, H., Marumo, F., Saito, Y.: Acta Crystallogr. *B28,* 1935 (1972)
88. Thöm, V. J., Boeyens, J. C. A., McDougall, G. J., Hancock, R. D.: J. Am. Chem. Soc. *106,* 3198 (1984)
89. Martin, L. Y., De Hayes, L. J., Zompa, L. J., Busch, D. H.: J. Am. Chem. Soc. *96,* 4047 (1974)
90. Bancroft, D. P., Boeyens, J. C. A.: 1985, to be published
91. Badger, R. M.: J. Chem. Phys. *2,* 128 (1934)
92. Boeyens, J. C. A.: Inorg. Chem. 1985, in press
93. Cotton, F. A., Walton, R. A.: Multiple Bonds between Metal Atoms, Wiley, New York 1982
94. Hambley, T. W., Hawkins, C. J., Palmer, J. A., Snow, M. R.: Aust. J. Chem. *34,* 45 (1981)
95. Boeyens, J. C. A.: Speculations Sci. Techn. *6,* 323 (1983)
96. Shull, H.: in: Physical Chemistry. An advanced treatise (eds. Eyring, H., Henderson, D.), *Vol. 5,* Academic Press, New York 1970
97. Daudel, R.: in: Ref. 98, 1980
98. Becker, P.: Electron and Magnetization Densities in Molecules and Crystals, NATO ASI Series B – Physics *48* (1980)
99. Hohenberg, P., Kohn, W.: Phys. Rev. *136B,* 864 (1964)
100. Kohn, W., Sham, L. J.: Phys. Rev. *140A,* 1133 (1965)
101. Perdew, J. P., Parr, R. G., Levy, M., Balding, J. L.: Phys. Rev. Lett. *49,* 1691 (1982)
102. Bransden, B. H., Joachain, C. J.: Physics of Atoms and Molecules, Longman, London 1983
103. Boeyens, J. C. A.: S. Afr. J. Chem. *33,* 63 (1980)
104. Boeyens, J. C. A.: S. Afr. J. Chem. *33,* 14 (1980)
105. Boeyens, J. C. A.: S. Afr. J. Chem. *33,* 66 (1980)
106. Boeyens, J. C. A.: J. Cryst. Spectr. Res. *12,* 245 (1982)
107. Boeyens, J. C. A., Ledwidge, D. J.: Inorg. Chem. *22,* 3587 (1983)
108. Hirshfeld, F. L.: in: Ref. 98, 1980
109. Boeyens, J. C. A.: J. S. Afr. Chem. Inst. *26,* 94 (1973)
110. Pretorius, J. A., Boeyens, J. C. A.: J. Inorg. Nucl. Chem. *40,* 407 (1978)
111. Ohwada, K.: J. Chem. Phys. *72,* 1 (1980)
112. Jahn, H. A., Teller, E.: Proc. Roy. Soc. London *A164,* 220 (1937)
113. Davydov, A. S.: Quantum Mechanics, 2nd ed., translated by D. Ter Haar, Pergamon, Oxford 1965
114. Bersuker, I. B., Polinger, V. Z.: Adv. Quantum Chem. *15,* 85 (1982)

115. Bentley, J., Stewart, R. F.: Acta Crystallogr. *A30,* 60 (1974)
116. Groenewegen, P. P. M., Zeevalkink, J., Feil, D.: Acta Crystallogr. *A27,* 487 (1971)
117. Coppens, P., Stevens, E. D.: Adv. Quantum Chem. *10,* 1 (1977)
118. Dunitz, J. D., Seiler, P.: J. Am. Chem. Soc. *105,* 7056 (1983)
119. Berlin, T.: J. Chem. Phys. *19,* 208 (1951)
120. Ermler, W. C., Kern, C. W.: J. Chem. Phys. *55,* 4851 (1971)
121. Hirshfeld, F. L.: Acta Crystallogr. *B40,* 484 (1984)
122. Smuts, J. C.: Holism and Evolution, Macmillan, London 1927

Electrostatics and Biological Systems

Shuk-Ching Tam and Robert J. P. Williams

Inorganic Chemistry Laboratory, South Parks Road, Oxford OX1 3QR, U.K.

The understanding of electrostatics including interactions of ions with the surfaces of proteins is not far advanced. There are two major problems – the mobility of the surface and the difficulty of handling electrostatic interactions in water. First this article analyses the electrostatic problems of ion-association in water. It begins with a consideration of small molecule/small molecule (anion/cation) binding observing that we are only able to proceed in an empirical way despite much theoretical analysis. The constraints are due to the difficulty of knowing how charged ions match one another in shape and how water screens the interactions. The description of ion binding to polyelectrolytes is shown to suffer from the same problems, and they are only intensified when we examine two or three dimensional surfaces interacting with ions. Empirically, that is through the analysis of experimental observations, we begin to see that we need a description of the energy of dynamic patches of mosaics of charge and hydrophobic areas in order to understand the binding of proteins to other ions and to one another. This is a very difficult task. It is probably the same problem as that of cell/cell interaction.

I. Introduction

It is clear that the folding of many biological polymers, their subsequent organisation and activities, and even the shapes of biological membranes are dependent in considerable part upon electrostatic interactions between their charged surfaces. This interaction must be such that it overcomes competition from small inorganic cations or anions which are also present within cellular systems. (In some cases the folding is assisted in fact by the binding of these small ions but this is not usual.) The major competing intracellular inorganic cations are K^+ (and some Na^+) at 10^{-1} M and Mg^{2+} which is present at 10^{-3} M while outside the cell Na^+ is present at 10^{-1} M and both Ca^{2+} and Mg^{2+} exceed 10^{-3} M. The small anions in biological solutions which are in high concentration and which might bind surfaces strongly include ATP^{4-}, HPO_4^{2-}, many phosphorylated sugars, and many carboxylated molecules which may well exceed 10^{-3} M in concentration in cells. Outside cells, the major anions are Cl^- and SO_4^{2-}. Most charges on polymers might be expected to be screened by these small charged ions and molecules. The question arises then as to how binding specificity can be generated by the organic polymer surfaces since at first sight electrostatics is merely a general attractive force and effects of the high concentrations of the small ions should be overwhelmingly important. In this paper we use empirical data to show that discrimination in the electrostatic terms can develop and that one form of selection between anion and cation partners arises from structural requirements, i.e. repulsive forces, while another develops from the matching of patterns of local charges on surfaces, which we shall call spatial charge matching. We start this examination from an analysis of the binding data for electrostatically bound complexes formed between small spherical inorganic anions and cations, move to the binding of small organic anions and cations and the binding of these small organic ions to inorganic ions, then to the binding of small ions to biopolymer surfaces, and finally to some remarks about biopolymer/biopolymer interactions.

II. Ionic Strength Effects

We shall not analyse in detail the general theories of electrolyte solutions which start from the theory of Debye and Hückel[1]. The simplest form of electrostatic interaction is that between point charges and is basically a Coulombic interaction. The Debye-Hückel theory for the electrostatic interaction between ions applies the Poisson-Boltzmann equation to the point-charge system in a medium of uniform dielectric constant and when expressed in terms of the potential ψ (for the condition $e\psi < kT$): –

$$\psi(r) = \frac{\varepsilon z_i \exp(-\varkappa r)}{Dr}$$

$\psi(r)$ = potential at a distance r from the ion

ε = electronic charge

z_i = charge on the ion

D = dielectric constant

$$\varkappa = \left\{\frac{4\,\pi\varepsilon^2 N}{DkT}\sum_i c_i z_i^2\right\}^{1/2}$$

$\frac{1}{2}\sum_i c_i z_i^2 = I$ = ionic strength, N = Avogadro Number

The expression for $\psi(r)$ can be applied to describe the distribution of the counter-ions around the ion (e.g. to find the Debye-length) or to evaluate the screening of ions from other ions in the solution. In essence this theory is not about ion-ion binding but it shows that as salt concentrations increase any given ion is stabilised in a solution by an electrostatic free energy of interaction with a self-imposed ion atmosphere. Here the stabilisation by salt of the solution $(\Delta G) = RT\log_e\gamma_\pm$ (where $\log_e\gamma_\pm = -A|z_1z_2|\sqrt{I}$) of the *free* ion is *independent* of ion size and A is a constant.

The equation was modified to take into account ion volumes, for example by Onsager[2)], but there remains virtually no selective free energy of interaction amongst ions but for charge differences[3)]. Real solutions and especially the biological solutions of interest here show very marked selectivity of interaction at given charge. This could be due to two factors: either the electrostatic theory in itself is incorrect for some ions or the assumption that electrostatic forces are the only ones to be utilised is incorrect. It turns out that both the electrostatic analysis and the assumption are not adequate. The first point becomes clear after a consideration of the binding of small ions to one another following Bjerrum's analysis[4)].

Before passing on note that Na^+, K^+ and Cl^- contribute largely to the above general salt background in biology and are of little consequence in binding to small or large organic molecules of opposite sign. The divalent ions are very different.

III. Ion-Pair Formation

For divalent ions strong deviation from Debye-Hückel behaviour is apparent and experimental data are then interpreted in terms of associated species. As an example to show the effect of charge on the formation of ion pairs, experimentally determined 1 : 1 binding constants of a range of simple anions to cations, extrapolated to ionic strength zero, are given in Table I. The data are taken from standard compilations. An approach to the understanding of the binding in these and in other ion pairs can be based on a very simple model, treating the ions as spherical ions in a medium of uniform dielectric constant D, as has been developed by Bjerrum[4)]. His equation for K, the 1 : 1 binding constant of A^+B^-, is

$$K = \frac{4\,\pi N}{1000}\left\{\frac{|z_1z_2|\varepsilon^2}{DkT}\right\}^3 \int_2^b y^{-4}\exp(y)\cdot dy \tag{1}$$

$$\text{where}\quad y = \frac{|z_1z_2|\varepsilon^2}{DrkT} \quad\text{and}\quad b = \frac{|z_1z_2|\varepsilon^2}{DakT}$$

where the symbols have the same meaning as before and a = distance of closest approach for the ions. Using this equation we can plot log K against $|z_1z_2|$ calculated for various

Table I. Representative stability constants of inorganic ion pairs, $\log_{10}(K_{I=0}/dm^3\ mol^{-1})$ at 25 °C

Cation	Anion	Charge product	$\log_{10}K$	Ref. (below)
$[Co(NH_3)_5OH_2]^{3+}$	$[Fe(CN)_6]^{4-}$	12	3.17[a]	1
$[Co(NH_3)_5py]^{3+}$	$[Fe(CN)_6]^{4-}$	12	3.38[b]	2
$[Co(NH_3)_6]^{3+}$	$[PO_3]_4^{4-}$	12	5.74	3
$[La(H_2O)_n]^{3+}$	$[Fe(CN)_6]^{3-}$	9	3.70	4
$[Co(NH_3)_5L']^{3+}$	$[Fe(CN)_5L]^{3-}$	9		
L′ = 4-phenylpyridine	L = 4,4′-bipyridine		3.02[c]	5
L′ = 4-phenylpyridine	L = 1,2 bis(4-pyridylethane)		2.88[d]	5
$[Ca(H_2O)_n]^{2+}$	$[Fe(CN)_6]^{4-}$	8	3.63	6
$[Ca(H_2O)_n]^{2+}$	$[Fe(CN)_6]^{3-}$	6	2.83	7
$[Co(NH_3)_6]^{3+}$	SO_4^{2-}	6	3.56	8
$[Co(NH_3)_6]^{3+}$	WO_4^{2-}	6	2.59	9
$[Ca(H_2O)_n]^{2+}$	SO_4^{2-}	4	2.19	8
$[Co(NH_3)_5NO_2]^{2+}$	SO_4^{2-}	4	2.69	10
$[Co(NH_3)_6]^{3+}$	Cl^-	3	1.85 ± 0.2	this work
$[Co(en)_3]^{3+}$	Cl^-	3	1.90 ± 0.2	this work
$[Co(en)_3]^{3+}$	ClO_4^-	3	1.15	11
$[K(H_2O)_n]^+$	SO_4^{2-}	2	0.85	12
$[Ca(H_2O)_n]^{2+}$	NO_3^-	2	0.68	13

[a] $I = 0.1$; when corrected to $I = 0$ using $\log_{10} f_\pm = -0.509|z_1z_2|[I^{1/2}/(1 + I^{1/2}) - 0.3\,I]$, $\log_{10} K = 5.70$
[b] $I = 0.1$; when corrected to $I = 0$ using $\log_{10} f_\pm = -0.509|z_1z_2|[I^{1/2}/(1 + I^{1/2}) - 0.3\,I]$, $\log_{10} K = 5.95$
[c] $I \approx 0.077$; when corrected to $I = 0$ using $\log_{10} f_\pm = -0.509|z_1z_2|[I^{1/2}/(1 + I^{1/2}) - 0.3\,I]$, $\log_{10} K = 4.8$
[d] $I \approx 0.077$; when corrected to $I = 0$ using $\log_{10} f_\pm = -0.509|z_1z_2|[I^{1/2}/(1 + I^{1/2}) - 0.3\,I]$, $\log_{10} K = 4.64$

1. Gaswick, D., Haim, A.: J. Am. Chem. Soc., 1971, *93,* 7347
2. Miralles, A. J., Armstrong, R. E., Haim, A.: J. Am. Chem. Soc., 1977, *99,* 1416
3. Monk, C. B.: J. Chem. Soc., 1952, 1317
4. Stampfli, R., Choppin, G. R.: J. Inorg. Nucl. Chem., 1972, *34,* 205
5. Gaus, P. L., Villanueva, J. L.: J. Am. Chem. Soc., 1980, *102,* 1934
6. Hanania, G. I. H., Israelian, S. A.: J. Solution Chem., 1974, *3,* 57
7. Gibby, C. W., Monk, C. B.: Trans. Faraday Soc., 1952, *48,* 632
8. Work, J. B.: Inorg. Synth., 1946, *2,* 221
9. Lyness, W., Hemmes, P.: J. Inorg. Nucl. Chem., 1973, *35,* 1392
10. Ueno, M.: Rev. Phys. Chem. Jpn., 1973, *43,* 33
11. Kaneko, H., Wada, N.: J. Solution Chem., 1978, *7,* 19
12. Helgeson, H. C.: Am. J. Sci., 1969, *267,* 729
13. Fedorov, V. A., Robov, A. M., Shmylko, I. I., Idanoskii, V. V., Mironov, V. E.: Zh. Neorg. Khim., 1969, *19,* 1746

assumed distances of closest approach, *a,* and we can compare prediction with the experimental results. Figure 1 shows that for the systems of highly localised spherical charge densities described here, the results agree with prediction, for values of *a* in the expected range for ion-pair contact[3)].

When we examine Bjerrum's theory closely however we see that it has used exactly the same assumption about electrostatic interactions as the Debye-Hückel theory. In fact the only new feature is the way in which space around an ion is described and it is this arbitrary division of space which has defined ion-pairing. In Bjerrum's theory any two

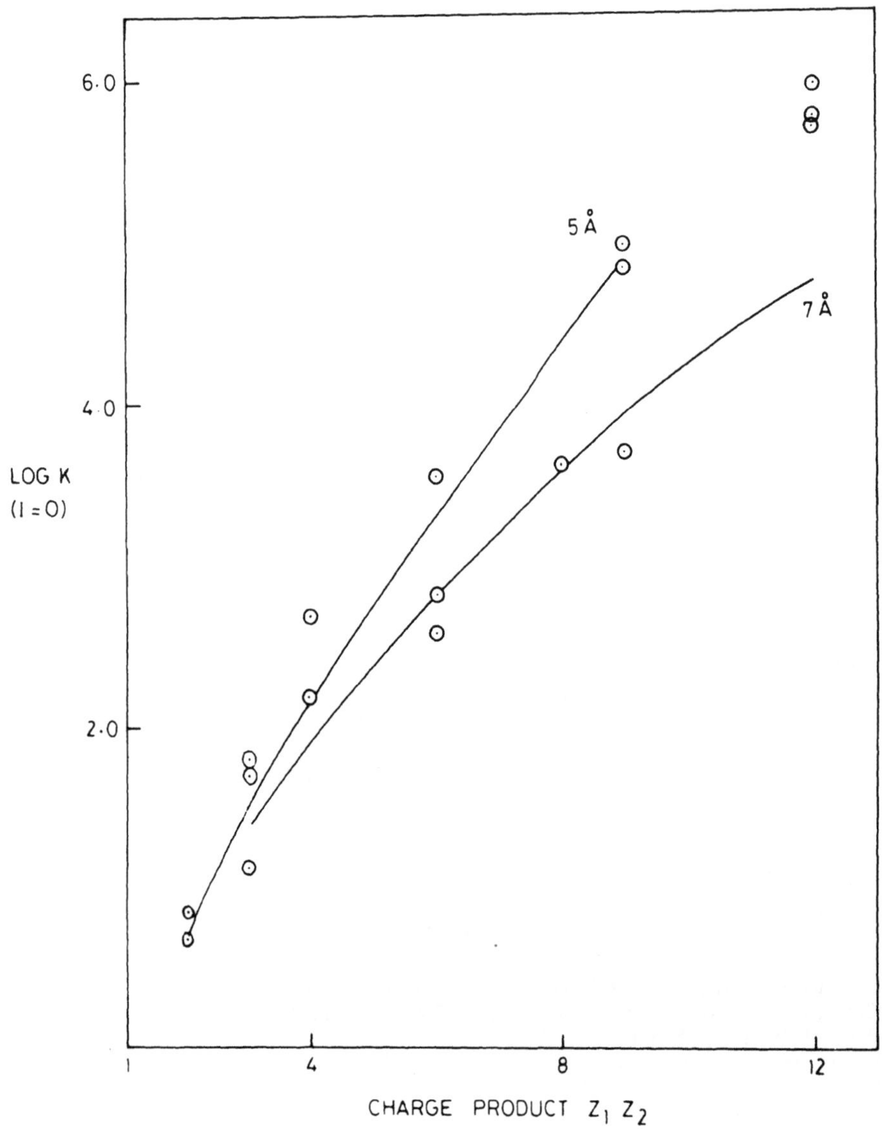

Fig. 1. The relationship between binding constant, log K, and charge product according to Bjerrum theory *(full lines)* for two distances of closest approach. The *circles* give experimental points from Table I

ions from those separated by a distance such that the electrostatic energy between them is equal to 2 kT up to a limiting distance, *a,* which is the distance of closest approach, are said to be paired. Now although the data in Table I and many data for other ions do indicate that ion-pair formation in real systems is approximately related to charge as shown by Fig. 1 there is also a peculiar variation with crystal ionic radii at fixed charge. These deviations become more apparent the more complicated the ions, Table II[3,5,6]. Sometimes the larger and sometimes the smaller cation has the bigger binding constant to an anion and Bjerrum theory takes this into account by giving a variable *a* for a given cation. *a* is therefore not an actual size (structure) but an empirical fitting thermodynamic parameter. It is sometimes obtained from measurements of log K and called the effective ionic radius in solution in an alternative procedure to the use of estimated *a* to evaluate the binding constant. We would prefer a theory which used known ionic radii from crystals and went on to describe binding constants. The reasons for wanting such a treatment follow from the observation that much of the data does not give agreement with Bjerrum's approach unless values of *a* are used arbitrarily. The discussion which follows will show that the observations can be better explained by a molecular picture of ionic interactions including, as an essential feature, the interaction with water molecules and by a consideration of other forces such as those involved in repulsion. Unfortunately these relevant features have not yet been described by equations.

Further comparison, Fig. 2, of Mg^{2+}, Ca^{2+} and Co^{3+} complexes of some more complicated anions shows that for these cations there are selective features amongst the ion-pair binding constants which are surprisingly quite strong and which can not be explained using Bjerrum's treatment[7]. The anions of Fig. 2 are built on a relatively rigid frame. The selectivity appears also in the data in Fig. 3 which is for the binding of anions with a high internal mobility. We have chosen the three simple cations since from crystal radii considerations, the bare calcium ion is larger than the bare magnesium ion, when calcium/magnesium comparison should reveal clearly the effects of changes of bare cation size, while $[Co(NH_3)_6]^{3+}$ is the simplest *inert* complex ion with the same radius as $[Mg(H_2O)_6]^{2+}$. It should reveal selectivity between very large cations with different charges or between a very large cation and a hydrated cation of lower charge which can be either a simple ion or a hydrated ion. In fact as Fig. 2 shows there are indeed selectivity effects since the order of binding constants between Ca^{2+} and Mg^{2+} can change and $[Co(NH_3)_6]^{3+}$ is obviously bound for example by maleate considerably more strongly than by oxalate, which is not true for the magnesium cation. How can we explain such observations?

The Figs. 4 and 5 show more extreme selectivity when organic cations are compared with inorganic cations[7]. As the length of the alkyl chain between two $-NH_3^+$ or $-NR_3^+$ groups is increased the binding to long chain α, ω-dicarboxylate anions increases while

Table II. Comparison of Ca^{2+} and Mg^{2+} binding constants

	Binding constant log K_{ML} (dm^3 mol^{-1})	
Ligand	Mg^{2+}	Ca^{2+}
Glycine	3.4	1.4
Imidodiacetate	2.9	2.6
Nitrilotriacetate	5.3	6.4
EDTA	8.9	10.7
EGTA	5.4	10.7
Acetate	0.8	0.7
Malonate	2.8	2.5
Citrate	3.2	4.8

See Williams[5,6] for further details.

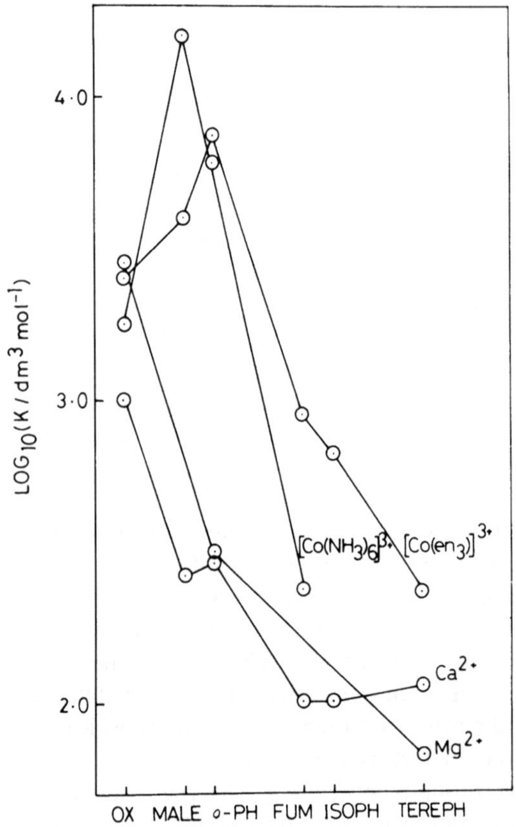

Fig. 2. A plot of binding constants for a variety of inorganic cations with rigid dianions. The dianions are ordered following increasing separation of negative charge, oxalate, maleate, ortho-phthalate, fumarate meta(iso)phthalate and terephthalate

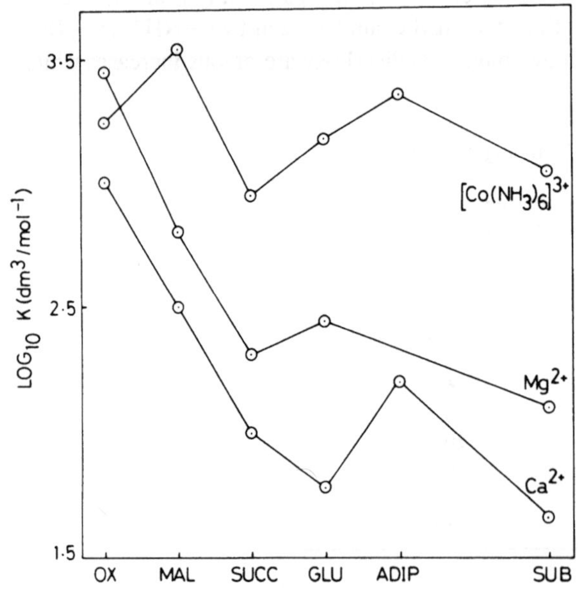

Fig. 3. A plot of binding constants for a variety of inorganic cations with flexible anions in order of charge separation, $^{-}O_2C$-$(CH_2)_nCO_2^-$ where *n* goes from zero to five

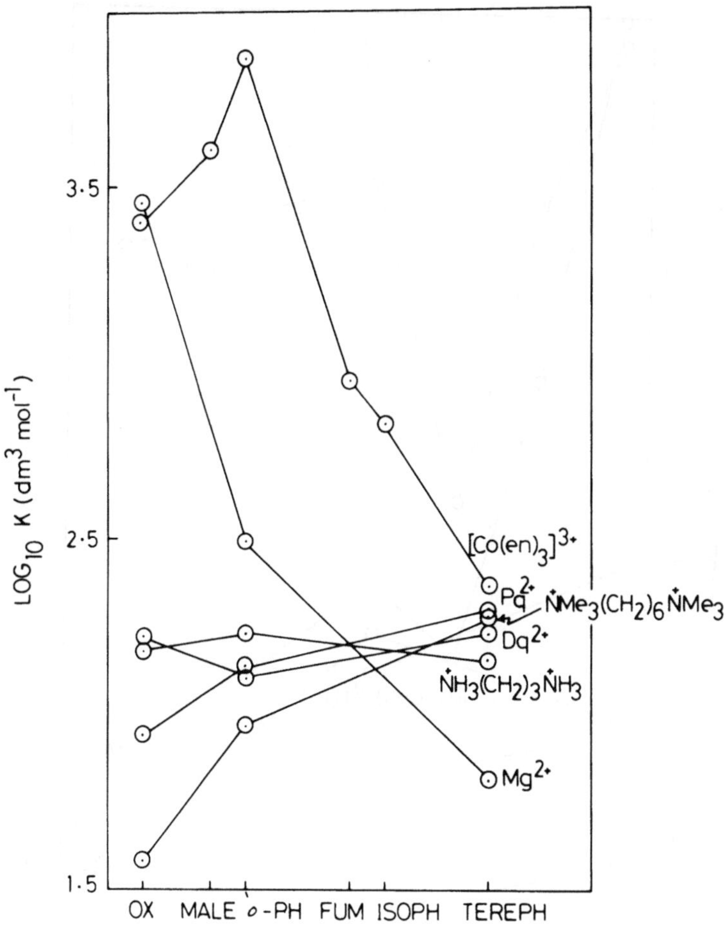

Fig. 4. A plot of binding constants for a variety of inorganic cations with rigid anions in order of charge separation. The organic cations are a mixture of flexible and rigid molecules, Pq is paraquat and Dq is diquat

the binding of the inorganic cations Mg^{2+} and Ca^{2+} along the same series of anions decreases. In the series the charge distribution of the inorganic cations Ca^{2+} and Mg^{2+} becomes a poorer and poorer match for the organic anion charge distribution but this is much less true for the large ion $[Co(en)_3]^{3+}$ around which the anions can wrap. The organic cations of larger charge separation show increased binding as the matching of charge distribution to that of the α,ω-dicarboxylate anions improves. No matter whether these selectivity changes occur through strictly electrostatic forces, through H-bonding or through the additional Van der Waals (hydrophobic) interactions between alkyl chains or all of the above, we can see that widely spaced negative charges do not have a high affinity for small cations but they do bind quite well to widely spaced positive charges which we shall call spatial charge matching. The same features are also present when we compare different anion bindings e.g. phosphates to a series of cations (Tam, Wilson and Williams to be published). We return first to the nature of simple spherical ions analysing the observations using a molecular model for the medium, water.

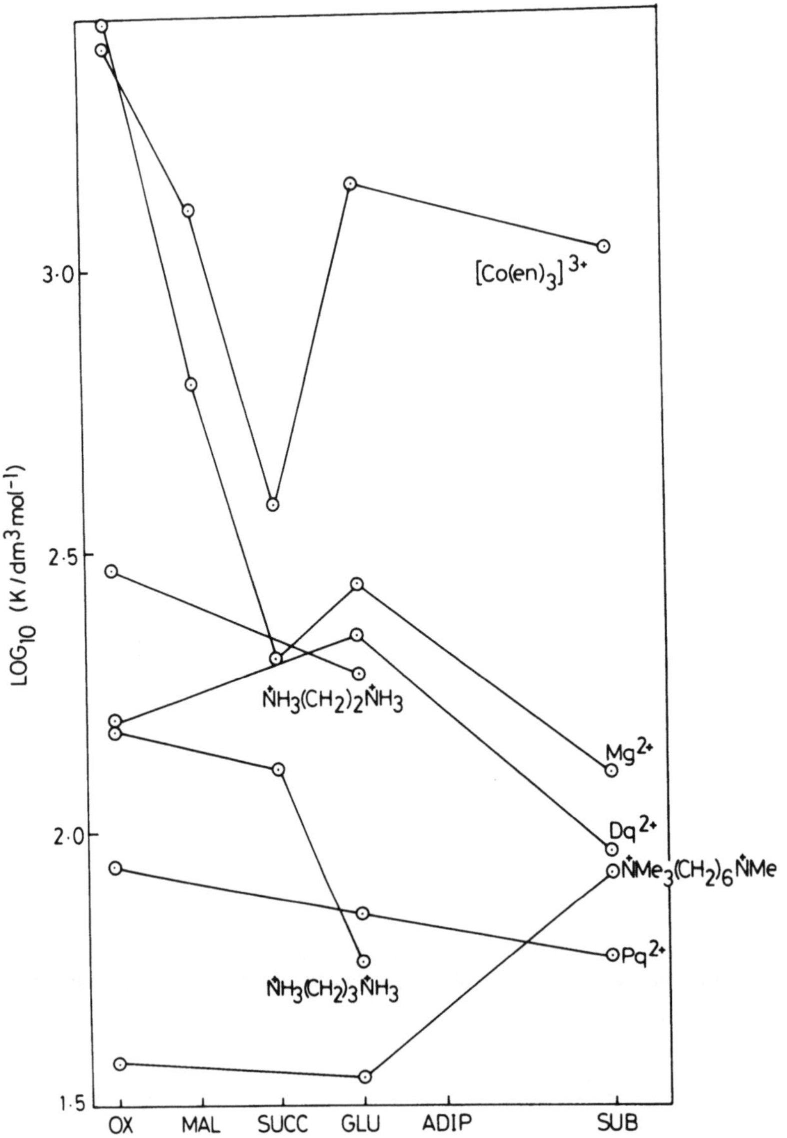

Fig. 5. A plot of binding constants for a variety of organic cations with flexible anions in order of increasing charge separation. The organic cations are a mixture of flexible and rigid molecules

A. The Concept of Ion Size: Inorganic Cations

It is convenient to allocate sizes to ions but as shown by Phillips and Williams[8], sizes of ions (or molecules) are derived from studies of solids in which the ion occupies a volume which is a function of the lattice forces. As is well-known the apparent sizes of ions changes from those in octahedral to those in tetrahedral anion holes. The commonly given ionic radii are those found in octahedral sites in oxides. The volume each of the ions occupies is decided by optimisation of the lattice energy or more generally, the optimisation of environment energy. There will be a small change in the apparent size of say Mg^{2+} relative to Ca^{2+} as we go from octahedral holes in MF_2 to octahedral holes in MCl_2 because of the greater restriction placed upon the packing of the chloride ions (the larger anion) around the magnesium ion (the smaller cation). However although this change in apparent radii (apparent since they are deduced from inter-ion distances) is small it means that the equilibrium

$$CaCl_2 + MgF_2 \rightleftarrows MgCl_2 + CaF_2$$

is considerably biased to the left. The energy of this bias may be but a few kilocalories yet in a chemical sense this is a considerable selectivity in log K. In effect repulsion between anionic members of the coordination sphere is contributing to log K. In aqueous solutions this becomes important since we always study a competition between water, which is a small group in the coordination sphere (compare fluoride), and anions which are nearly all bulkier (compare chloride). The repulsion between water and the anions and between water and water in the coordination sphere then has a large influence on log K. Similar considerations apply to the description of anion hydration. The theory of Bjerrum has only a radial field electrostatic attraction and no repulsive force except an excluded volume of radius a. (We see immediately that the size of a in fact is partly a reflection of the repulsion between ligands including bound water.) In part this is related to the problem of the concept in Bjerrum theory of water as a medium of uniform dielectric when in fact it is molecular. The general analysis of these effects of ion size has been treated for many years under the title of radius ratio effects. We now start to use "real" structural parameters in our analysis.

B. Spatial Radius Ratio Effects

Pauling pointed out that cations, being smaller than anions, had to fit into holes generated by close-packing of the anions in order to make solid lattices[9]. He observed that the hole sizes arising from different packings of anions were different and that cations of different sizes selected different holes e.g. in spinels and silicates. Williams[5] went on to consider the effect of this selectivity on the solubility of salts. The equilibrium is

$$M^{n+}(\text{hydrate}) + X^{n-}(\text{hydrate}) \rightleftarrows MX(\text{hydrate}) + m\, H_2O$$

Apart from the free energy gain of the water on the right it is clear that the difference in electrostatic energy between the product and the reactant (ions of charge z^{n+} or z^{n-} and radius r) is related to a difference between individual ion hydration energies and lattice energies[8].

$$\Delta G \propto C\left\{\frac{z^{n+}}{r_M} + \frac{z^{n-}}{r_X}\right\} - B\left\{\frac{z^{n+}z^{n-}}{d_{MX}}\right\} \qquad (2)$$

where d_{MX} of the latter $= r_M + r_X$ to a first approximation and C and B are constants. By using C and B as fitting constants we are taking into account repulsions approximately.

Simple examination of this function for a fixed r_X shows that at either very large or very small values of r_M the first term, the hydration energy, dominates. In a somewhat obvious way the lattice is most stable somewhere near the equality $r_M = r_X$. Detailed analysis allowing for $d_{MX} \neq r_M + r_X$ only reduces the importance of the lattice term. This means that ions of approximately equal size pack together best. Phillips and Williams[8)] showed how the solubility of salts could be discussed in detail using Eq. (2) and went on to demonstrate that even the residual structural hydration of ions in lattices in the most stable form of MX(hydrate)$_n$ was decided by the radius ratio of anion and cation. As the anion gets larger relative to the cation there is an increasing probability that a hydrated cation not the bare cation will come out of solution with a given anion and that the hydrated salts are then the most insoluble e.g.

$BaSO_4$, $SrSO_4$, but $CaSO_4 \cdot H_2O$ and $MgSO_4 \cdot 7\,H_2O$

which is to be contrasted with the series with a small anion.

$Ba(OH)_2 \cdot 2\,H_2O$, but $Sr(OH)_2$, $Ca(OH)_2$, $Mg(OH)_2$, and $Be_nO(OH)_m$.

The data obviously reflect the drive toward matching of sizes of ions. These observations show that the considerations of Bjerrum are likely to fail when we deal with real ion-pairing since the ease with which an anion can come close to a cation is decided not by the electrostatic energy z_1z_2/Dd where D is the bulk dielectric constant and d is the real distance between them (not a) but is controlled in part by the degree to which small molecular water is a preferred ligand. In other words we need a quite different model in which the stereochemical properties (size and shape) of the solvent molecules, here water, as well as of the anion and cation are taken into acount. It is the good packing of the small water molecules around magnesium ions (in the lattice) which gives $Mg(H_2O)_6^{2+}$ a higher stability in association with SO_4^{2-} (outer sphere binding) than Mg^{2+} (inner sphere binding). At the same time the reverse case is true for Ba^{2+} association with SO_4^{2-} compared with association by $Ba(H_2O)_8^{2+}$. The selectivity shown between hydrated and non-hydrated ion associations in crystals is based on the peculiarly small size of water. Small cations prefer to pack with small anions or dipolar molecules e.g. water[5–8)].

Now we shall consider the effect of radius ratio on the formation of ion pairs in solution[5, 8)]. There is a close parallel between the above analyses of solubility and that of the reaction for complex ion or ion pair formation

$$M^{n+}(\text{hydrate}) + X^{n-}(\text{hydrate}) \rightleftarrows [M^{n+} \cdot X^{n-}]_y H_2O \text{ or } [M^{n+}(H_2O)_y \cdot X^{n-}] + m\,H_2O$$

The form of the equilibrium energy for formation of the complex is the same as that in Eq. (2) but the constant B is different. In solution we expect therefore that size factors, of ions and water, will be important and that in some complexes it will be the hydrated ion which dominates to give species $M(H_2O)_nX$ whereas in others the bare ion will dominate to give $MX(H_2O)_m$. The difficulty in proving the point in solution lies in recognising the species. We must go to spectroscopic studies in order to recognise the species but the central point is that ion-pairs always have structure dependent energies as well as radii dependent energies.

C. Observed First and Second Coordination Sphere of Cations: Structure

A very important consideration in electrostatics as seen above is the relative importance of the first and the second coordination sphere around a cation[10]. The regular packing of ions or water around a central sphere gives rise to four coordinate tetrahedra, six coordinate octahedra and to an eight coordinate cube but five and seven coordinate packings are rare. One of the reasons for the absence of these coordination numbers is the difficulty of packing 5- and 7-coordinate systems. In solution this problem still exists in terms of the internal structures of the coordination spheres. The choice between 4, 6 and 8 (or any other) coordination is based of course on the best possible single set of M–X bonds given X–X repulsion and water attraction and repulsion. The first goes with $1/d_{M-X}$ and the second goes with a much higher power of d_{X-X}, d_{H_2O-X} and $d_{H_2O-H_2O}$. The constraint is the radius ratio of M and X. Our first expectation is that as M increases in size relative to X we should find higher coordination numbers. The simple rule is observed to be correct but some further effects are unexpected as can be shown by the study of crystalline hydrates. (In passing note that charged headgroups of biological polymers are normally quite large.)

The general rule works well in hydrated salts with small cations such as Be(4), Al(6), Cr(6), Mg(6), coordination number in parentheses, but fails in the hydrated salts of Na^+, K^+, Ca^{2+} and La^{3+} i.e. all ions of larger radius than 0.9 Å. In the crystalline salt hydrates of these larger ions, the cations form *irregular* structures in which the bond lengths, bond angles and even the number of oxygen atoms around the central ion seems to be almost arbitrary. Some typical structures are given in Tables III to V. The central field of the cation can no longer be dominant. Instead the interaction between water molecules in the

Table III. Some typical calcium salt structures

Salt	Ca(II) coordination no.	Ca–O distances (nm) Min.	Max.	Ref.
$CaHPO_4 \cdot 2\,H_2O$	8	0.244	0.282	(a)
$Ca(H_2PO_4)_2 \cdot H_2O$	8	0.230	0.274	(b)
Ca 1,3-disphosphoryl-	6	0.226	0.236	(c)
imidazole	7	0.227	0.278	(c)
Ca dipicolinate $\cdot$ 3 H_2O	8	0.236	0.257	(d)
$CaNa(H_2PO_2)_3$	6	0.231	0.233	(e)
Ca tartrate $\cdot$ 4 H_2O	8	0.239	0.254	(f)
$Ca(C_6H_9O_7)_2 \cdot 2\,H_2O$	8	0.239	0.247	(g)

(a) Beevers, C. A. (1958). Acta Cryst., *II,* 273–277.
(b) MacLennan, G. & Beevers, C. A. (1956). Acta Cryst., *9,* 187–190.
(c) Beard, L. N. & Lenhert, P. G. (1968). Acta Cryst., *24B,* 1529–1539.
(d) Strahs, G. & Dickerson, R. E. (1968). Acta Cryst., *24B,* 571–578.
(e) Matsuzaki, T. & Iitaka, Y. (1969). Acta Cryst., *25B,* 1932–1938.
(f) Ambady, G. K. (1968). Acta Cryst., *24B,* 1548–1557.
(g) Balchin, A. A. & Carlisle, C. H. (1965). Acta Cryst., *19,* 103–111.

N. B. When calcium is six-coordinate the structures may be much more regular.

Table IV. Some Ca^{3+} structures of small molecules of biological interest

	Coordination no.	Ca–O distances (nm) Min.	Max.	Ref.
Ca^{2+} thymididylate	7	0.230	0.265	(a)
Ca^{2+} diphosphonate	8	0.240	0.260	(b)
Ca^{2+} galactose	8	0.235	0.255	(c)
Ca^{2+} blephavismin	7	0.235	0.245	(d)
Ca^{2+} trehalose	7	0.235	0.255	(e)
Ca^{2+} arabonate	8	0.245	0.250	(f)

(a) Trueblood, K. N., Horn, P. & Luzzati, V. (1961). Acta Cryst., *14,* 965–982.
(b) Uchtman, V. A. (1972). J. Phys. Chem., *76,* 1304–1310.
(c) Cook, W. J. & Bugg, C. E. (1973). J. Am. Chem. Soc., *95,* 6442–6447.
(d) Kubota, T., Tokoroyama, T., Tsukuda, Y., Koyama, H. & Miyake, A. (1973). Science, Wash., *179,* 400–402.
(e) Cook, W. J. & Bugg, C. E. (1973). Carbohydrate Res., *31,* 265–275.
(f) Furberg, S. & Helland, S. (1962). Acta Chem. Scand., *16,* 2373–2383.

Table V. Some typical magnesium salt structures

Salt	Mg(II) coordination no.	Mg–O distances (nm) Min.	Max.	Ref.
Mg hexa-antipyrine · ClO_4	6	0.206(Ca, 0.230)	0.212	(a)
$Mg(C_4H \cdot O)_4 \cdot Br_2$	6	0.216(Oxygens, 4)		(b)
$MgS_2O_3 \cdot 6\,H_2O$	6	0.205	0.212	(c)
$Mg \cdot SO_4 \cdot H_2O$	6	0.204	0.209	(d)
$Mg(HPO_3) \cdot 6\,H_2O$	6	0.200	0.212	(e)
$Mg_2P_2O_7$	6	0.200	0.211	(f)
$Mg_2P_2O_7$	6	0.200	0.211	(f)
$Mg(CH_3CO_2)_2 \cdot 4\,H_2O$	6	0.200	0.210	(g)

(a) Vijayan, M. & Viswamitra, M. A. (1968). Acta Cryst., *24B,* 1067–1076.
(b) Perucaud, M. & Le Bihan, M.-T. (1968). Acta Cryst., *24B,* 1502–1505.
(c) Baggio, S., Amzel, L. M. & Becka, L. N. (1969). Acta Cryst., *25B,* 2650–2653.
(d) Baur, W. H. (1962). Acta Cryst., *15,* 815–826.
(e) Corbridge, D. E. C. (1956). Acta Cryst., *9,* 991–994.
(f) Lukaszewicz, K. (1961). Roczn. Chem., *35,* 31–37.
(g) Shankar, J., Khubchandani, P. G. & Padmanabhan, V. M. (1957). Proc. Indian Acad. Sci., *45,* A, 117–119.

Note. Several magnesium hexahydrate salts have also been examined but crystal structure data on lower hydrates of magnesium are rare.

primary coordination sphere and the next layers of anions and water molecules must have increased relatively to the point that the observed geometry around the cation is not a central field effect but a cooperative compromise with demands of outer sphere and solvent structure. The overall picture is then that small ions prefer regular structures, octahedra and tetrahedra, but larger ions (0.95 Å) accept almost any irregular pattern. The conclusion is that for small cations a few very good central bonds give the most stability while for larger cations the central ion coordination sphere is adaptable in order to give an optimal overall energy including the second and higher coordination spheres. This distinction is a product of a great variety of interactions such as H-bonding between solvent molecules and steric factors. Its consequences in solution are dramatic in that

(1) Mg^{2+}, Al^{3+} show relative slow exchange of ligands
(2) Ca^{2+}, La^{3+}, Na^+, K^+ show fast ligand exchange
(3) Ligands on Ca^{2+}, La^{3+}, Na^+ and K^+ move easily – fluctional complexes – but do not do so on Mg^{2+}, Al^{3+}.
(4) Mg^{2+}, Al^{3+} seek a well-defined octahedral hole
(5) Ca^{2+}, Na^+, K^+ and La^{3+} have little demand on bond angle, bond length or coordination number in the first coordination sphere
(6) The ligands for Mg^{2+} and Al^{3+} are the strong field ligands e.g. RO^- or H_2O, not carbonyl, ether or alcohol
(7) The ligands for Na^+, K^+, Ca^{2+} and La^{3+} can be such groups as carbonyl or ether in place of anions or water presumably because the hydration of anions or of water itself overwhelms their electrostatic interaction with the larger cations.
(8) The ability to form many bonds of irregular geometry makes ions such as calcium, but not magnesium, excellent cross-linking agents.

These facts create a structural, thermodynamic and kinetic selectivity of effect in biological systems such that Mg^{2+} especially is in a different category of ion from Ca^{2+}, Na^+ and K^+. The general calcium trigger in biology is based on these properties.

[We have already made the point that most of the charged headgroups of biological molecules are relatively large e.g. RNH_3^+, $(RO)_2PO_2^-$ and so on. It follows that for all such groups, no fixed hydration pattern or geometries are expected. The packing of charged groups of biopolymers and small charged molecules is very easily open to adjustment and will depend critically on the neighbouring ions. We return to such organic based-groups later.]

D. Binding Constants for Ion-Pairs Again

That ion-pair formation is controlled by the same radius ratio considerations as those which decide lattice energy, solubility and the form of hydration of the most insoluble salt can be further illustrated by the selectivity of ion-pairing, log K, within series with on the one hand small anions and on the other large anions e.g.

for F^-, OH^-: $Be^{2+} > Mg^{2+} > Ca^{2+} > Sr^{2+} > Ba^{2+}$

and for SO_4^{2-}: $Ba^{2+} > Sr^{2+} > Ca^{2+} > Mg^{2+}$

These series parallel series of insolubility and of hydration in lattices. (They are interpreted as changes in the size parameter, *a,* by Bjerrum's treatment.) There is every reason to suppose that the same factors control the formation of ion pairs on the surfaces of proteins, membranes or polysaccharides where distinct differences are found between $R\text{–}OSO_3^-$ and $R\text{–}CO_2^-$ groups in their orders of binding of the above cations.

It follows from the above considerations of radius ratio effects that three situations can bring about tight binding (inner sphere) between highly and oppositely charged ions.

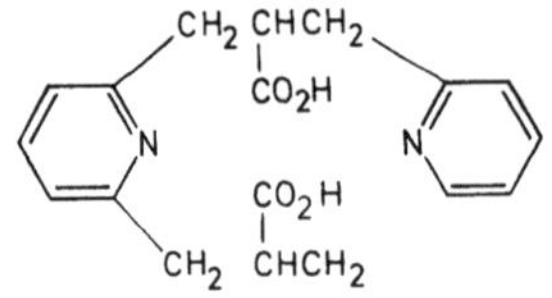

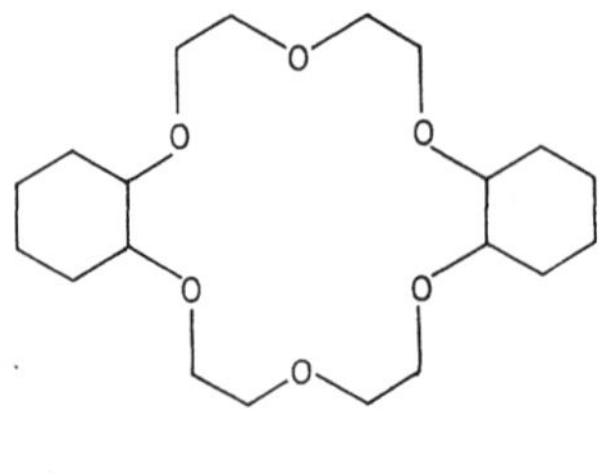

Fig. 6. Two types of ring chelate for selective binding. The top chelate was proposed by Williams in 1952[11]. The bottom chelate was developed by Pedersen[12], a similar biological ligand was found by Pressman and a series of subsequent workers, Lehn[13], Cram[14] and others have developed many organic syntheses of ligands of this kind

The first situation is that of a highly localised charge density e.g. $Mg^{2+} + P_2O_7^{4-}$. In effect Bjerrum's treatment caters for this case, Fig. 1. The second possibility is where there is a low repulsive constraint between ligands as for example when a fixed hole, created by a multidentate ligand anion, a cryptate, is a very good fit for the cation in question. One way of creating such holes is by organic frame synthesis when repulsion between the chelating centres has no effect since the donor groups are restrained by covalent links, Williams[11], Pedersen[12], Lehn[13], Cram[14] and others, Fig. 6. In biological systems Mg^{2+} is known to be bound very well in a small hole in a few instances (chlorophyll) while a larger cation or anion can also be selected for by a larger cavity. We shall see later that the presence of carbonyl, ether and alcohol oxygen donors in cavities on the surfaces of proteins has an additional effect upon the selective binding of calcium over magnesium. This arises from a ligand-chemistry factor. The third case is the coming together of large, very poorly hydrated, ions e.g. Ba^{2+} and NO_3^-. In a paper[7] discussing the data in Figs. 2 and 3 we have shown how these considerations can be used to explain the selective effects in the more complex systems of Figs. 4 and 5.

E. Selectivity Amongst Inorganic Anions

When describing the chemistry of *spherical* cations in terms of simple electrostatics, it will appear as if our understanding is at least adequate, see above. The selectivity on the basis of charge and radius can probably be explained so long as proper consideration is given to structural factors including those of the hydration spheres although there is no quantitative treatment of the latter. The situation with anions is more complex. Firstly there are very few simple spherical anions like F^-, Cl^-, Br^- and I^- and even their binding selectivity does not appear to depend on electrostatics alone. At one end of the series H-bonding to fluoride is very strong so that in water F^-, apart from giving HF_2^-, can not be represented by a negative charge and a radius unless we restrict discussion to sites which have no H-bond capability e.g. inorganic lattices. At the other end of the series, I^-, a different extra energy term arises in that I^- is highly polarisable and Van der Waals forces other than H-bonds become important. For this reason I^- can associate with cations which are also polarisable or polarising and may form complexes of a quite unusual kind e.g. I_3^-, which are surprisingly stable in water. Fluoride is very different from iodide in ways we did not see on going from lithium to caesium cations.

A different problem arises if we attempt to study the effect of increase of anion charge. Species such as O^{2-}, S^{2-} and so on have such a powerful affinity for protons that they and even OH^- and SH^- are of little real interest in aqueous solution chemistry at pH = 7. They do not enter into electrostatic ion pairs with most of the free cations of biological interest Na^+, K^+, Mg^{2+} and Ca^{2+} and they do not compete effectively for the surface positively charged holes of proteins, nucleic acids or lipids with anions such as phosphate. We note the great contrast of O^{2-} and S^{2-} with Ca^{2+} and Mg^{2+}. The special association of the proton with O^{2-}, OH^-, S^{2-} and SH^- as opposed to the weak hydroxide ion association with Ca^{2+} and Mg^{2+} arises from the special chemistry of the proton and cannot be described by electrostatic forces alone.

The common anionic species of biological systems are derived then from oxy-anions of lower pK_a in particular SO_4^{2-} and HPO_4^{2-} together with their carbon-bound forms $R_2PO_4^-$, RPO_4^{2-}, RSO_4^- and RCO_2^-. SO_4^{2-} and HPO_4^{2-} are the simple anions which dominate biochemistry together with Cl^- and small quantities of NO_3^-, SeO_4^{2-}, $B(OH)_4^-$, Br^-, I^-, F^- and some complicated anions such as SCN^-. Consider first the competition between oxyanions of the same charge and shape based upon XO_4 tetrahedra. The main anions are $ROSO_3^-$, SO_4^{2-}, $(RO)_2PO_2^-$ and $ROPO_3^{2-}$. Note that they are all large and we do not expect well-defined coordination geometry. The binding to cations of these anions of similar charge and size is quite different due apparently to the difference in electron affinity of the central atom Cl > S > P. This is seen in the pK_a values of $ClO_4^- < HSO_4^- < H_2PO_4^-$ and $SO_4^{2-} < HPO_4^{2-}$. The pK_a values reflect the charge density on the oxygen in the unit XO_4. In fact $R{-}OSO_3^-$ is a very poor coordinating centre not only for protons but for all cations and $(RO)_2PO_2^-$ is a very modest one. Sulphate ion pairs are less strong than $ROPO_3^{2-}$ ion pairs. It also means that RCO_2^- is closely similar to the weak acid di-anion (HPO_4^{2-}) and very different from the above tetrahedral oxy mono-anions. We have an order of electrostatic binding as follows

$$RO(PO_3)^{2-} > RCO_2^- > (RO)_2PO_2^- > ROSO_3^- .$$

Charge and size alone are not useful descriptions of these complicated anions. We shall inspect this series again later.

We saw earlier that as charge density of anions increases so affinity for all cations increases but selectivity moves from larger to smaller cations. Barium sulphate but magnesium phosphate are the most insoluble of the alkali earth salts. Note that biological precipitates (minerals) are found with the pairings $Ca_2(OH)PO_4$, $CaCO_3$, CaC_2O_4 but $SrSO_4$ and $BaSO_4$. The insoluble oxides of biology are formed only with very small cations e.g. SiO_2, Fe_2O_3. Oxide is a very small anion.

F. Amines as Cations: Complexity of Shape and Size

We turn next to complicated organic cationic species of a wider variety of size and shape, Fig. 4 and 5. There are two major types of amines in biology – those derived from ammonia and those from guanidine. The cation headgroups are large, at least as large as Rb^+ but they are also quite unlike simple cations in that they have a shape. NH_4^+ is tetrahedral and the guanidinium ion is roughly trigonal $\left[-NH-C\begin{smallmatrix}NH_2\\NH_2\end{smallmatrix}\right]^+$. The positive charge density is on the hydrogen atoms and is grossly spread away from the central atom. Their structures do not differ greatly from the anions such as ClO_4^- (tetrahedral) and NO_3^- (trigonal). Again we do not expect charge to define functional capability. How do these peculiarities affect their properties and what happens as we move from NH_4^+ to the still larger cation $N(CH_3)_4^+$ which is incorporated into compounds such as choline?

The localisation of positive charge on the protons of NH_4^+ groups has the immediate consequence that anion binding is re-inforced by hydrogen-bonding. The hydrogen bonds are stronger the higher the pK_a of an anion so that we expect ammonium ions to have a greater affinity for oxy-anions of high pK_a e.g. $RPO_4^{2-} > RCO_2^- > ClO_4^-$. We cannot demonstrate this clearly in the oxy-anion salts but it is shown to be the case in the series of halides, Fig. 7, where ammonium fluoride is clearly anomolous. The implication is that if H-bonds are strong there will be a special affinity for some anion sites such that NH_4^+ or RNH_3^+ will bind but Rb^+, or K^+ or $N(CH_3)_4^+$ will not bind as strongly. We expect

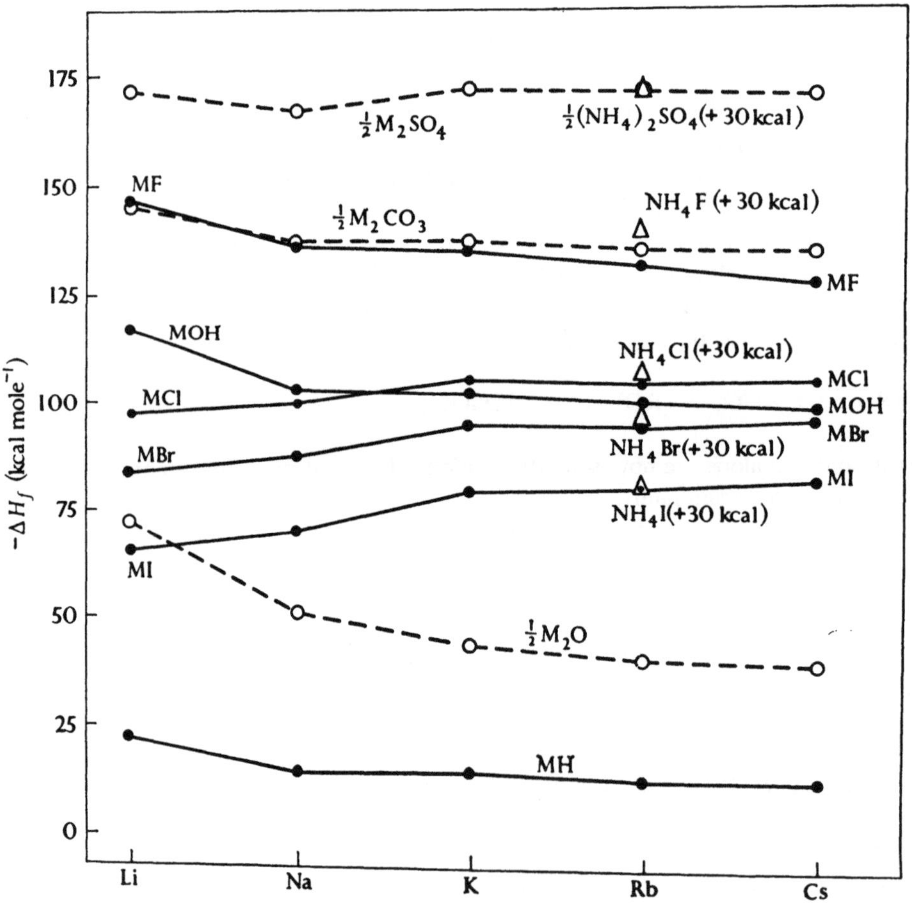

Fig. 7. The heat of formation of various 1:1 salts. Note how NH_4F is peculiarly stabilised by hydrogen bonding

that NH_4^+ will go most easily to RPO_4^{2-} sites and will not be much associated with $(RO)_2PO_2^-$ or RSO_4^- sites. There is the added selectivity factor as shown by Lehn and co-workers[13] that the binding site can be the correct shape (not only the correct size) for the H-bonds of NH_4^+ [13–18].

Guanidinium is a huge cation, much larger than Cs^+. On grounds of size it will seek out a large hole. It allows and forms strong H-bonds in crystals but it has a peculiar geometry fitting the two oxygens of an anion such as phosphate. The guanidium group, because of its size, is readily dehydrated, cf. Cs^+, and in fact this cation is a *denaturing* agent (like urea) for proteins while NH_4^+ is used as a *crystallisation* or *lattice stabilising* cation.

The tetramethylammonium cation has a large radius but does not form H-bonds. Despite its poor interaction with water it also interacts extremely poorly with inorganic oxy anions. It is then just an agent for carrying positive charge comparable with RSO_4^-. Methylation is a way of reducing interaction with anions in water. We expect that all these cations can behave quite selectivity through their structures. Many of these features are reflected in the data of Fig. 4 and 5 as discussed in Ref. 7. It should be remembered that solvation of all these organic anions and cations is likely to be highly irregular and variable as for calcium, sodium and potassium.

G. Matching of Anions and Cations

We have observed that the greatest selectivity for atomic spherical ions is generated by the construction by synthesis of an organic framework hole which exactly matches one ionic size[11–18]. Water is the preferred ligand for smaller ions whereas repulsion excludes larger ions than the one which gives a perfect fit. Since the ions carry charge, the binding to the hole is assisted by charges of opposite sign carried by the framework and often as direct coordinating centres. Extending the notion of the guest molecule from simple (atomic) ions such as Ca^{2+} or Cl^- to more complex ions such as NH_4^+ and SO_4^{2-} and then to organic cations and anions means that selectivity can increase as the host cavity is constructed more and more to reciprocate the different chemical functionalities of the guest[7]. If we proceed along these directions we see that we arrive at the general notion of a lock (host) and key (guest) which has been the language of enzyme/substrate or receptor/agonist chemistry for nearly a century. The framework does not have to be closed as for a hole but can be open-sided, a groove, and somewhat flexible to allow rapid on/off rates. This changes the description from the lock and key fit to the induced fit but the limitations, repulsions and attractions, of fitting dominate still. Selectivity will be relaxed as we increase flexibility or reduce functionality until we are back to the flexible organic ions discussed at the beginning of this article. We shall need to know if the discussion of selectivity needs any amplification as we now increase the size of both anion and cation so that they are both large open organic frameworks like DNA and proteins.

H. Summary of Electrostatic Binding of Small Ions

At this stage we see that in aqueous solutions electrostatic inter-actions cannot be described by reference to the radial field forces from ions, e.g. Bjerrum theory. Any electrostatic binding to the surface of proteins whether it be of water molecules, of small ions of charge $\leqslant 3+$, of other protein surfaces or any other biopolymer surfaces will depend not only upon the properties of the individual groups but also upon the matching of the charge patterns and on the steric constraints arising from the packing of molecular entities. We must also consider local charge on exposed atoms if an ion has shape e.g. SO_4^{2-}. Finally without invoking covalence we have to examine the effect of H-bonding. When all these terms are put together we expect selectivity of interaction between anions/cations and water to appear but we will not be able to explain our observations without a detailed examination of average structure if this is possible. We must remember that in solution electrostatic forces do not necessarily constrain structure closely and the dynamics of structures may partially hide structure itself. We need a statistical thermodynamic molecular approach and that based upon a bulk dielectric constant is bound to fail.

We can now examine the question of selectivity further and with respect to biology by turning to the general equation for the effect of an ion in a reaction system

$$\text{Effect varies as } K[M] \cdot \exp \psi/RT \cdot [\text{Structural Dynamics}] \tag{3}$$

The equation states that the effects due to different cations may depend upon the free ion concentrations, the equilibrium binding constants of each ion and the electric field potential. However even if these three terms were exactly the same the *effect* of binding could be quite different since the structures (i.e. the position of the atoms in space) and the dynamics of the structures can be quite different. It is not usual to suppose that these structure dependent parameters are of great consequence in electrostatic systems but Table VI and the enormous difference in the functions of calcium and magnesium in biology indicate that electrostatic interactions can be highly selective even when the first three terms of Eq. (3) are closely similar.

The situation of small organic anions and cations in both non-biological and biological structures is similar but more complicated since they are structurally more complex and have a lower surface charge density yet they can form strong H-bonds.

Table VI. Some anions and cations present in biology

(a) Simple Ions		
Cations	Anions	Interactions
Na^+, K^+, Mg^{2+}, Ca^{2+}	Cl^-, RCO_2^-, $ROPO_3^{2-}$	Mg · ATP
(b) Linear Polymers		
Cations	Anions	
Polyamines, Histones	RNA, DNA, Acidic Proteins	Histone · DNA
(c) Two-dimensional Surfaces		
Cations	Anions	
None	Membrane	Ca · membranes
	Phospholipids	Ca · Trigger proteins

Once more we stress that special structures reflect particular thermodynamics of interaction. We must be very careful when we describe on the basis of simple models the relative affinities of anions for cations.

IV. The Real Extent of Binding Between Small Biological Molecules and Ions

In the introduction we have pointed to the concentrations of ions in biological fluids. Na^+ and K^+ are roughly 0.1 M, Ca^{2+} and Mg^{2+} are roughly 10^{-3} M, Cl^- is 0.1 M but SO_4^{2-} and HPO_4^{2-} and their derivatives are 10^{-3} M. The concentration of enzyme substrates rarely exceeds 10^{-3} M (e.g. the carboxylated anions of the molecules of the citric acid cycle) and local charges rarely exceed two. We can draw some very simple conclusions about ion association in biology before we turn to polymers.

(1) The predominant small inorganic and organic cations and anions will not associate with one another. The exceptions are the polyphosphates e.g. adenosine triphosphate (ATP) and pyrophosphate. They will always be bound to Mg^{2+} (in cells). Since the surface charge densities on all biological polymers are low we expect no direct binding of small ions unless the polymer forms a cavity i.e. a purpose built structure.

(2) The biological interest in electrostatics now depends on the disposition of charges on organic frameworks. The presence of these frameworks not only introduces steric repulsion factors, charge-matching and H-bonding capacities but also hydrophobic forces. For small frameworks all binding remains small, Figs. 4 and 5. We must turn to a more detailed analysis of frameworks. We shall always have to remember that these frameworks have some structural mobility both locally and globally. Binding is a statistical thermodynamic problem and not just a structural one. The retention of mobility at all stages is also a necessity if on/off reactions are to be fast.

A. *Organic Frameworks*

When charges are built on an organic framework, we have to consider the distribution of the charges on these frameworks, either in a one or a higher dimensional array. The binding selectivity resulting from the matching of charge distribution of small oppositely charged ions was illustrated in Fig. 4 and 5. The selectivity is observed for both rigid and mobile frameworks. It is not yet possible, however, to resolve quantitatively the different contributions (i.e., charge matching, steric repulsion, hydrogen-bonding and hydrophobic factors) to the observed selectivity. For the small molecules such as those shown in Fig. 4 and 5, contributions from hydrophobic interaction (water exclusion) may not be dominant. However, the hydrophobic factor becomes very important in biological structures in which the organic framework is usually extensive, e.g., lipid bilayers, proteins and DNA.

That the hydrophobic interaction increases with the extent of the organic framework is illustrated by the observation that the free energy (stabilisation) of the partition from water to hydrocarbons solvents is greater for tryptophan than for alanine. The dominant importance of hydrophobic interaction is also known by the formation of micelles in water. Here the hydrophobic interaction between the hydrocarbon tails is larger than the electrostatic repulsion at the head groups, e.g. in fatty acid soaps.

Therefore in the consideration of the stability of structures in a system, we have to look at not only the electrostatic but also the hydrophobic term. For example, the addition of salts such as Na_2SO_4 or $(NH_4)_2SO_4$ may not affect the ionic bindings in proteins so much as the addition of organic molecules e.g. urea or ions such as guanidinium. The addition of salts reduces electrostatic ion-pairing but enhances hydrophobic salting out. The two effects can balance one another. It is then the small organic molecules not salts which disrupt the biopolymer assembly the most. We are also familiar with the extraction of inorganic ions by amphiphilic reagents into organic media. The combination of electrostatic (head group/ion) interaction with hydrophobic (hydrocarbon tail/solvent) interaction is responsible for the stability of the inorganic ion pairs in the organic medium.

It is the interplay of electrostatic and hydrophobic interactions, together with other contributions, such as H-bonding, that are responsible for the properties and reactivities shown by charged biopolymers. We turn next to the detailed examination of the nature of biological frameworks.

We shall proceed as follows. At first we consider the general case of a chemically unspecified rigid frame with a spaced out set of charges on it. This is the simplest linear framework. We shall then ask what is the general effect of specifying chemically the charges and the charge separation on it and what is the effect of making the framework mobile. We shall consider real examples. We then turn to the same problem in two dimensions – a frame which is a plane of smeared charge with no chemical definition. We then ask the same questions as of the linear problem. Virtually all the surfaces will be negatively charged. Finally we look at more complex surfaces such as those of proteins where there is a mosaic of charges of different types. The hydrophobic interactions will be introduced from time to time to remind the reader of their presence.

V. Polyelectrolytes – Linear Frameworks

An electrostatic theory of polyelectrolyte solutions was developed over a period of years culminating in the work of Manning[19]. It is similar in some ways to the work of Bjerrum on ion-pairs. The theory describes the electrostatic interaction of an infinitely long rigid polyelectrolyte chain which is given a uniform charge density, $z_p\varepsilon/b = (\beta)$ where b is the distance which separates individual single charges e.g. any two adjacent anion phosphate diester groups of DNA. The interaction potential with a cation is then given by

$$u_{cp} = -z_c\varepsilon(2\beta/D)\ln r$$

where c represents the cation and p the polyanion, D is the bulk dielectric constant of water and r is the distance between the point charge cation and the rod polyanion which is assumed to be rigid. Note the form of the electrostatic energy is now logarithmic in distance r.

To obtain a sense of complex ion formation we must divide space, as Bjerrum did, around the polyelectrolyte. Calculation of the number of ions inside the volume from zero to a distance r_0 from the anion follows from the equation

$$A_c(r_0) = f(r_0) \int_0^{r_0} \exp(-u_{cp}(r)/kT) 2\pi r dr \qquad (4)$$

which is of similar form to the Bjerrum integral for point charges except that Bjerrum defined a volume within which ion-pairing was said to occur, i.e. between the distance, r_0, where $(2\,kT = \varepsilon^2|z_1z_2|/Dr_0)$, and the distance of closest approach, *a*, see page 107.

We now need to propose the volume element, i.e. the values of *r*, between which the cations are said to be paired with the polyelectrolyte anion. Curiously for high charge densities this proves to be unnecessary. If b is small (e.g. less than 7.135 Å for $z_c = z_p = 1$), i.e. the spacing between charges on the polyelectrolyte is small, it can be shown that the integral diverges when $\xi > 1$ where ξ is $\varepsilon^2/DkTb$. This means that an initial system of dissociated charges as described is not stable under any conditions. In other words given the above assumptions, at $\xi > 1$, the cations will cluster immediately around the anion until the charge density is reduced such that $\xi = 1$. In effect there are no free cations until this condition is met which is found to be when approximately 80% of all anion charges are neutralised. No attempt is then made to resolve the binding energy. The binding of ions to polyelectrolytes are examined by Manning in terms of a so-called *condensation*. Association constants are thus implied to be infinitely strong up to this degree of neutralisation. (Note that *"condensation"* is *not* of the polymer, i.e. condensation is here closer to a precipitation of cations on to a linear anion whereas a biochemist describes condensation as a collapse of an extended polymer to a denser state, e.g. a folding.) Condensation here is a kind of ion association but which is only dependent on the charge density of the *rigid* (assumed) polyanion and is independent of the size of the cation. The extent not the strength of condensation depends somewhat on the charge of the cation.

It is worth saying again what the above equations have generated. They show that for a large polymer which is represented by a rigid *line* of dense point negative charges and at all reasonable dilutions of this polymer and its co-cation, say down to 10^{-10} M cation, there is always the same degree of loose association of about 80% of the cations with the anions. There is then no DNA^{n-} where *n* is the number of phosphate anion charges in the presence of K^+ but only $DNA_n \cdot K_{0.8n}$. In real cells some of the K^+ could be replaced by Na^+ (about 10%) and some by Mg^{2+}. The un-neutralised charge of the polymer is expected to fluctuate all over the chain through fast movement of the cations and it interacts with the residual 20% of the cations weakly. Notice that the association is not defined structurally in any way. We turn to the competition from polyamines later.

Interesting though Manning's condensation is and although it has been used successfully in some cases, it is far from being the complete picture of small ion association with polyelectrolytes[20, 21]. We must consider also ion pairs of the polymer with cations (Bjerrum style) for polymers of lower anionic charge densities too and for local regions of high charge on the polymer. In this case the integral of Eq. (4) has to be solved for the case where the ion pair association energy is equal to or exceeds 2 kT. We partition the energy as in the simple Bjerrum theory and examine the equilibrium. As the charge density gets small we reach the lower limit of ion pair formation and pass into a region where only

Debye-Huckel theory applies with the potential modified for that of a polyelectrolyte. Remember too that no selective effect of different cations has been described so far.

It follows that it becomes important for us to note the following about a polyelectrolyte (a) is it rigid to a first approximation? (b) what is the spacing of the regular charges? (c) are there regions of high and low charge density? (d) Do polyelectrolytes bearing different groups of the same charge, e.g., RCO_2^-, $(RO)_2PO_2^-$ and $ROSO_3^-$ show equal binding properties? (e) What are the structures of these associations? Given certain conditions we may apply Manning theory. However, we need to ask the same questions about it as we asked of Bjerrum's theoretical treatment. As it stands it can not explain competition between cations of the same charge but of different size and shape. There are no parameters in the theory to deal with these properties. We are forced back to empirical observations before we go further. It will become very important to know if K^+, Na^+, Mg^{2+} or polyamines bind best to linear polyelectrolytes of different kinds. Binding by K^+, Na and Mg^{2+} may not prevent DNA expression but binding of polyamines may very well do so.

A. *Linear Polyelectrolytes in Real Situations*[22]

We have just asked whether the generalised line-charge representation of a polyelectrolyte is satisfactory since there are at least four major types of anion $ROSO_3^-$, $(RO)_2PO_2^-$, RCO_2^-, and $R\text{-}OPO_3^{2-}$, listed in order of increasing pK_a. We know from experience (see above) that for the isolated mono-anions their increase in pK_a is associated with an increase in ion-association with all cations. While sulphate esters act almost as point charges their selectivity order for cations is the inverse of that of RCO_2^- or of $R\text{-}OPO_3^{2-}$ i.e. for sulphates or sulphonates it is Large cations > Small cations. This is the basis of the two different orders of elution of cations from different chromatographic ion-exchange resins, e.g. in the separation of lanthanide or actinide elements. Now as well as a change of order there is an increase in ion-pair affinity from sulphate to carboxylate surfaces.

At this point we must note again that Manning's theory is for rigid polyelectrolyte rods. The polymers of biology are not rigid and a very important part of their functions is the change in folding which occurs on addition of different cations. A typical example is DNA. Biologists describe DNA as "condensing" in the presence of cations, Table VII.

Table VII. The Folding of DNA and RNA[23-25]

Cation	Folding
Simple Cations[a] e.g. Mg^{2+}	Little folding but crosslinking where the polymer generates local anion charge density, Fig. 9
Dynamic Polyamines of no preferred shape (spermine etc.)	Total collapse of DNA (RNA) into small volume (sperm, pollen, spores)
Partially shaped cationic polymers e.g. histones	Folding of DNA into units around the folded polycations as in nucleosomes, supercoils

[a] Only electrostatics involved

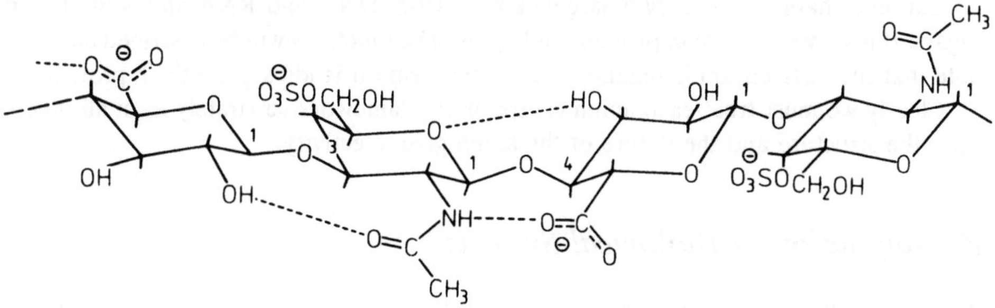

Fig. 8. The formula of a repeating sugar. In many polysaccharides of biological origin the charges are irregularly placed or only at chain terminals

The word condense has now two meanings – the collapse of cations into the surrounds of the rigid anion rod (polyelectrolyte theory) and the collapse of the extended forms of polymers in the presence of cations into a folded form (biochemists observation)[22]. Here we are interested in both. We have already made statements about individual anion affinities for cations. The best general statements we can make concerning the ability of different poly-anions to remain in an extended conformation in the presence of inorganic cations is that poly-sulphates will not collapse around cations under the normal conditions of biology, i.e. 0.1 M MX, and 0.005 M MX_2. We expect that polyphosphates will collapse under these conditions and the polycarboxylates and poly diester-phosphates will show intermediate behaviour. Moreover the degree of collapse depends selectively upon the counter-ions. This is of the greatest possible interest since the conformations of DNA and RNA as well as of membranes are then under the general control of electrostatics but with specific features. Particularly interesting are the situations in which inorganic cations compete with organic polycations, polyamines such as spermine, for DNA, Tables VI and VII. It is this competition which is part of DNA expression. [When we come to describe two-dimensional frame-works we shall again find that the condensed (folded) forms of membranes, e.g. the myelin sheath, depend on the cations present]. Remember too as specific folding occurs it generates the special cavities which further enhance selectivity. Empirically we know that the larger cations $K^+ > Na^+$, and $Ca^{2+} > Mg^{2+}$, cause the collapse of polymers much more effectively than smaller cations and this has been ascribed to cross-linking ability, page 117, i.e. an effective fitting together.

The disposition of charges on biological polymers is also intriguing. Poly-diester-phosphates, RNA and DNA, correspond quite closely to the Manning description of a polymer with regularly repeating charges. Polysulphates on sugar backbones, poly-saccharide polyelectrolytes, are quite different since although sulphation can be quite regular (Fig. 8) it is frequently found that the charges are very irregularly spaced. In the case of random coil linear proteins too the charge distribution is idiosynchratic and it is this fact that enables the use of protein charges to create a unique series of organised units. Chromogranin A could be an example. Finally we turn to mono-ester phosphate anions and note that this group is relatively rare and doubly charged. The introduction of dianionic phosphate and its removal from polymers are under metabolic control so that the anionic frameworks can be regulated inside cells. We shall see the importance of the introduction or removal of such a high local charge density later.

Cationic charge in biology is largely absent from DNA and RNA and is limited in sugars. It is very extensive in proteins and lipids. The functions will be described later but note that the distribution is regular in lipids but again it is idiosynchratic in proteins.

Clearly we must treat each anion centre on its framework separately from all others since the structure and the nature of the anion give selectivity.

B. Sulphation in Biological Systems

We start the discussion of the different use of different anion centres from the knowledge that different anions of the same charge type show different strengths of binding to cations and that they show different cation selectivity orders. General theories of binding based on charge alone are inadequate. We also know that theories based on ionic radii, and media of uniform dielectric constant fail to explain observations. We must proceed empirically and try to rationalise later.

From the study of sulphate ester complexes with H^+, M^+, M^{2+} or anions it appears highly likely that they are the most dissociated anion centres in biology. Their use, like that of perchlorate in electrochemistry, is therefore an effort to generate, through repulsion, chemically inert but physically valuable spread-out anion atmospheres. In the case of biology the most prominent use is in sulphated sugars on the outside of cell membranes. These anionic polymers are mobile and in no way completely folded though there are local structural elements, Fig. 8. Little or no binding is useful here since the anionic polymer then fills a large space and can act as a sensor for the presence of constricting objects, see below. However, binding can occur to organic polymers due to surface matchings, see later.

A second instructive case is myelin. Myelin contains a high concentration of sulphatides. We must conclude again that this is a design to avoid binding of any kind to simple cations. There must then be either no binding except to a specially constructed positively charged polymer. Now myelin forms a *compact* sheath around the nerve cell fibres and there must be a compact zone where the sulphatide is. We therefore presume that there must be a basic protein to assist the organisation. One such basic myelin protein is known and in isolation it too is a mobile protein of little ordered structure. The combination of sulphatide polymers and the basic protein could generate the collapsed structure of both. This would be an example of framework charge – matching obviously assisted by hydrophobic terms. (The hydrophobic term and H-bonds may dominate in specific recognition but this is not our concern in this article.)

It is here that we have to make a special note about biological systems. They are compartmental. The sulphated polymers, polysaccharides, are outside cells only. The reasons are two fold. Firstly sulphur compounds of high oxidation state such as sulphate are only stable outside the cell since they are open to reduction to RSH, that is to low oxidation state compounds, inside cells. This is not true of phosphate and indeed of carboxylate. Again sulphation is of hexoses which are only polymerised and attached to proteins outside cells. The hexoses are relatively hydrophobic but carry no hydrophobic side-chains. These polymers of hexoses have one other feature which is not found in intracellular polymers – they are often branched chains, see later. All of these factors drive polyhexose sulphates to remain as rather open three-dimensional networks of unbound anions, but they are exposed to the external solution which is high in calcium ions (with little or no binding) whereas the inside of the cell is not. Biology has evolved so that certain types of anion see only certain types of cation. Note how these factors affect Eq. (3).

C. Diester Phosphates

The diester phosphate head group is not very different from the sulphate group in that it has a low affinity for all simple cations. Being derived from phosphate it has two condensation links R–OPO–R in contrast with the terminal binding of sulphate. Diester phosphates are then part of linear polymers, contrast also carboxylate. They are mainly used in combination with long alkyl chains to form the basis of membranes or in combination with nucleosides to form DNA and RNA. The polymers have very hydrophobic groups so that internal hydrophobic forces generate particular polymer forms – bilayer membranes and double helices. (In the case of RNA more complex folding patterns arise.) All in all then these polymers have anionic centres which are sufficiently well spaced out that they are not able to bind to simple cations such as Mg^{2+} and Ca^{2+}, with any strength, Fig. 2 and 3, except when folding gives rise to cavities. There is a further check on the associative interactions on membrane surfaces in that the phosphate diester group usually carries a short mobile side chain which is cationic e.g. ethanolamine or choline which further screens the negative charge. Usually the membranes still carry some residual negative charge which is treated as a Gouy-Chapman layer, see below. The negative charges of DNA and RNA are not screened internally and are sufficiently densely distributed to be treated by the Manning-type theory. Such theories however neglect the more hydrophobic parts of these polymers together with the possibilities of charge matching mentioned earlier. The diester phosphates in RNA especially can be arranged to give specific or at least highly selective folds which are associated with cavities, compare proteins. A fine example is provided by t-RNA where the Mg^{2+} ion is associated with particular binding cavities[23]. The ion usually remains highly hydrated and can be replaced by such cations as $[Co(NH_3)_6]^{3+}$ but the cavity-fold cannot be maintained by polyamines. In t-RNA we have then the two equilibrium[23, 24]

Closed form + Cations ⇄ Closed form. Cation Complex

Open form + Cations ⇄ Open form. Cation Complex

The balance in such equilibria i.e. two complexes of different fold energy biased by the binding of cations depends upon the difference in fold free-energy, the sequence of the polymer, and the availability of the different cations. The *sequence* of t-RNA leads to the possibility of binding Mg^{2+} to the exclusion of organic amines because of the peculiarities of its fold. It is a single strand with double helical regions interrupted by special folding, Fig. 9.

Consider now the reactions of DNA which forms a continuous *double* helix but can super-coil, Table VII. The equilibria which can be considered are[25]

Super-Coil + Cation ⇄ Super Coil. Cation Complex

Double Helix + Cations ⇄ Double Helix. Cation Complex

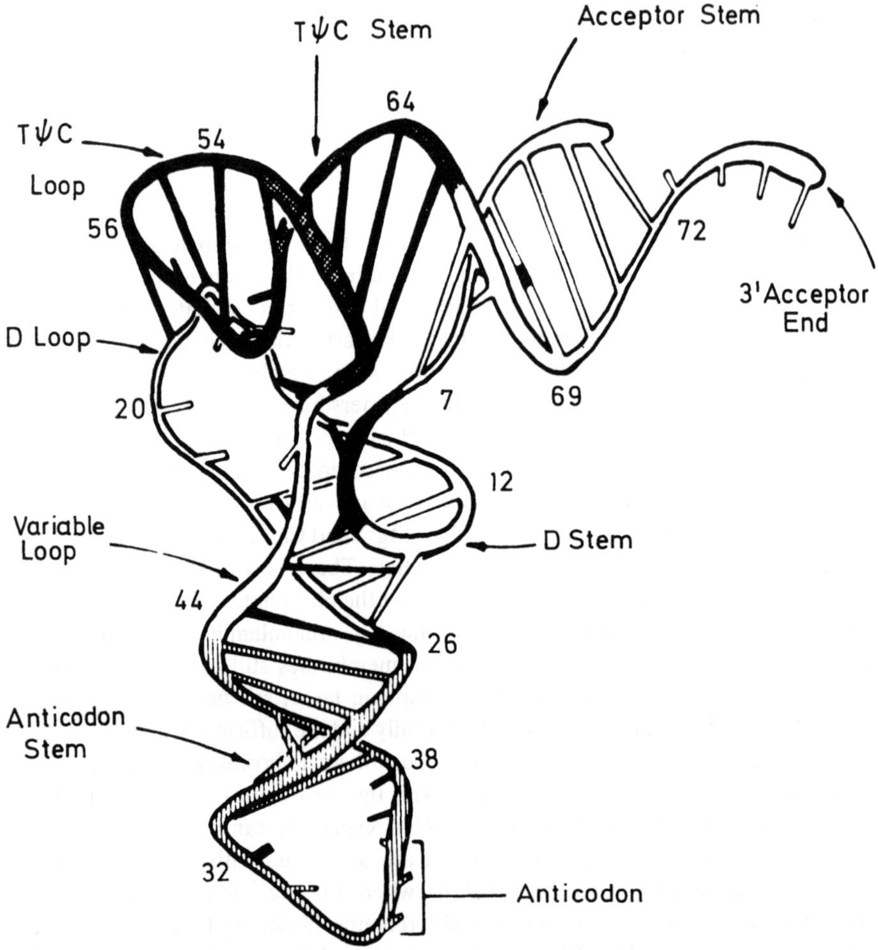

Fig. 9. The structure of t-RNA. The fold embraces and is stabilised by Mg^{2+} ions. In particular the tight turn from residue 8 to 11 is stabilised

The first types of cation to consider are again simple cations which now find no cavities in either the double helix or the super-coil, or various polyamines which can charge-match with the surface of the double helix or the super-coil. All the small cations can do is to bring about condensation (Manning) in a non-selective way but the polycations can cause specific foldings[23–25]. We know that biological systems have generated a series of polyamines (and basic proteins called histones, see later) which prevent the Manning type condensation by simple cations but which match the different "condensed" state of DNA in nucleosomes and in sperm and pollen "condensates". These folded states are not associated so much with small cations as with polyamines which have their own fold energy to contribute to the total stabilisation of the super-coil in various forms. Likewise we know that spermines are themselves virtually random coils in isolation, but are folded and associated with bacterial DNA especially in sporulation. Once again as with sulphated polymers the organic polyamines dominate the ion-pair interactions through charge matching. However, we must remember that these "structures" have very short lives and are constantly fluctuating.

Finally we refer to the distribution of di-ester phosphates and polyamines in biological compartments. One major group of diester phosphates are mostly in the nuclear compartment (DNA) and the ribosomes RNA but m-RNA and tRNA are more widely spread in the cytoplasm. All these polymers see 10^{-3} M Mg^{2+} but virtually no Ca^{2+}. The diester phosphates of membranes are not in a special compartment since they face every compartment. There are very few free di-ester phosphates in the external aqueous solutions of organisms. The polydiester phosphates bind to calcium on the external faces of membranes since only outside the cell is there sufficient calcium. The distribution is then very different from that of the sulphate-based anions.

D. Carboxylates

Unlike the lipids, the polysaccharides and the polynucleotides, the anionic groups of random coil, linear, proteins are almost invariably carboxylate groups. The carboxylic group is a much smaller group than the sulphates and the phosphates and the carboxylic group can be a stronger electrostatic binding group than the diester phosphates and sulphates. In synthetic polymers we expect carboxylates to form ion-pairs rather than ion atmospheres as for sulphates. (See Delville and Lazlo[21] for a discussion of the problem.) However, carboxylate groups in proteins are not regularly spaced. In fact it is very rare for carboxylate groups to appear regularly except in polysaccharides where they function somewhat like sulphate, Fig. 8, and also in the formation of special networks with the calcium ion. The role of carboxylate generally is more associated with folded polymers (see below). This is also true of the dianion phosphate mono-ester. Given that gene-controlled synthesis can lead to any dense or widely disposed series of carboxylates the selective uptake of inorganic or organic cations is under the control of highly specific systems.

E. Summary of Biological Linear Polyelectrolytes

The general low charge density and specific chemical nature of the anionic groups in sugars, DNA, RNA, membranes and many polyelectrolytes suggest that they avoid ion-pairing with atomic inorganic cations of biology. They are screened only by longer-range interaction of the Debye-Hückel kind as developed by Manning and Leary. This allows selective ion-pair through charge-matching and hydrophobicity with organic polyamines of various kinds. We note that rigid structures are uncommon and interactions are usually between highly mobile and mobilisables units so that static structural pictures could be very misleading. All the cations move easily over the surface of the anions so that rearrangements are kinetically fast. We expect, and all the model data given early show, that no simple treatment of electrostatics will explain these interactions.

We can envisage many kinds of open chain/chain association and also many collapsed states from any combination of segmented non-rigid linear polymers with other polymers or with ions. The interest lies in the specificity of these changes, both upon chemical character of the ions and on concentration changes. One interesting theoretical analysis of such systems is by Odijk for linear electrolytes[26]. We give the case of branched chain electrolytes later.

F. Polycations

The common polycations are low molecular weight polyamines such as spermine. There are also a number of very basic proteins, e.g. histones, but they are not extended rigid chain polyelectrolytes since they have a hydrophobic folded core. The distribution of positive charge in these proteins and in their polymers is then partially fixed on a three dimensional surface. The surface and the extreme tails will change with addition of counter-anions however so that specific shapes evolve based on the hydrophobic core. In this way nucleosomes are formed. We shall have to consider competition between different positive proteins for negative DNA, and the stereochemical results of the competition, in the presence of specific competing effects of small ions which can fit local cavities, e.g. $Mg^{2+} \cdot$ tRNA, see Fig. 9. On the whole Manning's theory does not apply to biological polycations since none of the simple polycations could be said to be (infinitely) long.

VI. The Electrostatics of Rigid Surfaces (Two-Dimensional Frameworks)

The conventional approach to surfaces of regularly spaced charge (anionic) and of high charge density is to replace the point charge by a smeared charged surface and then to consider the effect of a potential (ψ) i.e. the Gouy-Chapman approach[27, 28]. Gouy solved the problem of the distribution of anions and cations close to the surface using the Boltzmann equations

$$c_+ = c_0 \exp(-\varepsilon\psi_0/RT)$$

$$c_- = c_0 \exp(+\varepsilon\psi_0/RT)$$

The solutions are exactly the same in form as in Debye-Hückel (and Manning) theory but ψ_0 takes a different form. The so-called Gouy-Chapman layer is then the ionic atmosphere of a planar surface. The development of equations for the thermodynamics of these layers follows from the use of the Boltzmann-Poisson equation. We shall not give expressions here but refer to Refs. 27 and 28. In principle we can extend this treatment to ion-pair formation on the surface of the membrane when we shall have a constant K for binding to some fixed charge but of course it now depends on the membrane potential through an equation of the general form

$$K = K_0 e^{\psi/RT}$$

where K_0 would be the association to a single site in the absence of a potential. ψ for a charged surface falls continuously with distance from the surface due to the binding of counter-ions. K_0 is then a simple ion-pair constant described earlier. Written in this way we can treat K_0 as if it were the interaction of the head group of the surfaces, say $-CO_2^-$ or $(RO)_2PO_2^-$ or $ROSO_3^-$, as observed for simple anion interaction with a cation in water, see above. We must notice however that the part of the equation in ψ contains no selectivity term due to anything but the charge of the cation so Na^+ behaves like K^+,

Ca^{2+} like Mg^{2+}, and Ln^{3+} like $[Co(NH_3)_6]^{3+}$. On the other hand K_0 is an empirical constant and contains radius and shape dependent selectivity. Some cations break through the hydration zone of some of the anionic charged surfaces and in fact this will alter the effect of ψ to some degree. We shall not concern ourselves with this effect here. We can see that in general the same problems will arise in the description of a plane of charges as in the description of a line of charges, see above, but a new problem arises – curvature which parallels folding of a linear polymer.

A. *Shapes of Membranes*

Curvature of a biological membrane is very usual. It is known to be dependent upon the asymmetry of distribution of both lipids and proteins, especially their charges, across the membrane but it is not treated by Gouy-Chapman theory. This asymmetry is only part of the force acting on the membrane since the strong asymmetries of potential and of specific cations (energised gradients) interact with the charge asymmetry due to the lipids and proteins. Where there is change of curvature there must be a force along as well as across a membrane. Particularly obvious are the changes of curvature of mitochondrial and chloroplast membranes on energisation. We return to this point later but clearly the effect of free polyamines and cation selectivity must be considered in the same way as for the foldings of DNA and RNA and protein particles. Curvature is an expression of structure in three dimensions and is an immediate reflection of specific local binding.

B. *The Integration of the Boltzmann-Poisson Equation*

Before we leave the discussion of the principles of electrostatics we should note that there are always two kinds of difficulty in the theory. One is the basic model and its limitations which we have stressed again and again. There are the following limitations: (a) No quantitative account of molecular solvent/ion interaction to explain association reactions involving small energy changes. (b) No satisfactory treatment of the dielectric constant of the solvent. The other kind of difficulty rests in the mathematics even when every simplification of the field forces is made. The integration of the Boltzmann/Poisson distributions are made only in the very dilute solution range for 1 : 1 salts. This is unsatisfactory of biological systems where charges are higher and are often more concentrated. Resort to computerised non-analytical solutions leaves the experimentalist puzzled as to the effects of varying the parameters he can control. These factors are common to ion-pairs, linear polyelectrolytes and membrane surfaces. We may be forced to a very empirical approach.

C. *Branched Chain Polyelectrolytes: Three Dimensional Charge Patterns*

A further problem arises when we consider (three-dimensional) branched chain polyelectrolytes which are quite different from linear rods or planar membranes since they are often very mobile. Almost invariably they are based on polysaccharides in biology. In the

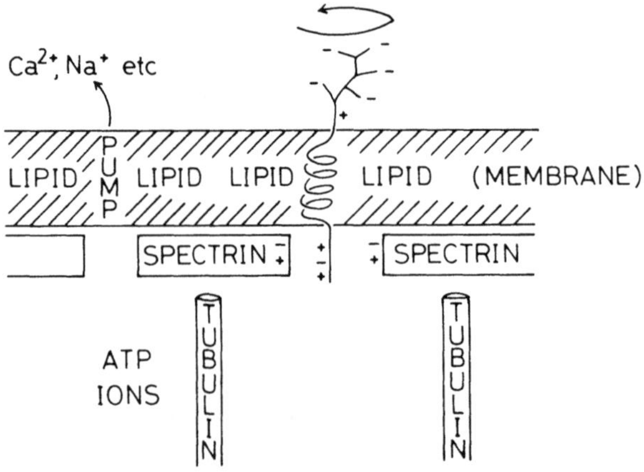

Fig. 10. The relationship between glycosylated membrane proteins and membranes. The polysaccharide is shown as a negatively charged branched chain filling much space. It is connected by a helical rod through the membrane. The polysaccharide can act as a receptor of information which is transmitted by the rod to the interior of the cell. The information is subsequently connected to cation movements. Many external signals e.g. of hormones may be processed in this manner[30)]

absence of cations they fill space as much as possible and have a very large configurational entropy, Fig. 10. On addition of counterions the structure may collapse to a very small volume. One way of causing collapse is to reduce (negative) charge repulsion just by adding salt. This follows from the Debye-Hückel ionic strength considerations. An alternative is to consider ion-pairing, say between the anions of the saccharide polyelectrolyte and calcium[29)]. We can then write

$$n\ Ca^{2+} + \text{polymer} \rightleftarrows Ca_n^{2+}\ \text{polymer (collapsed)}$$

At low concentrations the calcium just causes the polymer to contract in volume, curve 1 of Fig. 11 following a simple linear equilibrium binding. We stress calcium here since it brings about cross-linking most effectively in biology. There is a relationship between the polymer volume and the calcium ion concentration similar to that between volume and pressure in the expression for gases, PV = RT. In the latter equation,

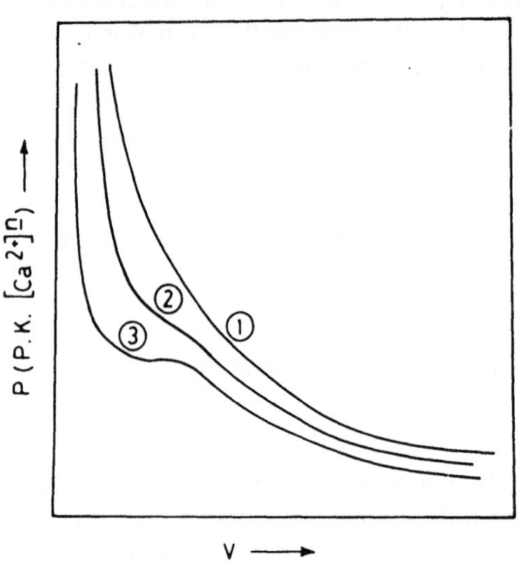

Fig. 11. The relationship between the volume filled by an expanded branched polysaccharide and the calcium ion concentration. *Curve 1* is for simple binding and *Curve 3* is for a condensational collapse (phase change) of the polymer[29)]

increasing the external pressure leads, at fixed temperature, to a decrease in volume. Analogous constants can be applied to the polymer in solution, but owing to the complexity of the system, compared with an ideal gas, additional features must be considered. The polymer volume increases with increasing negative charge, whereas pressure and gegen-ion effects – specifically calcium binding – will cause a reduction in volume. At fixed polymer concentration the equation can be written as:

$$P[\text{external}] \cdot [Ca^{2+}] \cdot V = RT$$

Now if the effects of calcium become cooperative then the relationship of V(volume) to $[Ca^{2+}]$ changes just as the $PV = RT$ expression changes at temperatures where a liquid can form, curve 3 of Fig. 11. In the case of calcium cooperativity it may be temperature changes or the structure of the polymer which brings about the change from curve 1 to curve 3 in Fig. 11. Be this as it may the overall effect is to connect small calcium changes to very great changes in volume as in phase transformations, curve 3. The constituent polysaccharides of cell walls have the potential to undergo such transformations. The external surfaces of membranes may behave in exactly the same way as the walls of cells either through the polysaccharides or through the lipids. Phase changes (sol to gel) are likely to be due to changes in calcium interactions in quite a number of cases. The interaction between a cell surface and an inert surface can also be dependent on "phase" changes as pictured in Fig. 10. (Note that "phase" changes are biological not Manning "condensations".) We have shown how these collapses can be used as sensors[30].

Now the model we have in mind of a total collapse of a random polymer on increasing salt or a binding cation has a parallel in the conformation changes of partially ordered polymers, see the next section. Here the change with salt or specific cations is not so gross but it can be looked upon in the same way in terms of the conventional $R \rightleftarrows T$ state transition or in the more general terms of one ensemble of conformers being *gradually* converted into another. We shall see that it is usual that when a loosely constructed protein such as calmodulin (negative) or a histone (positive) is bound by a set of opposite charges, calcium or DNA respectively, that the proteins change shape. The adjustments of shape are made easy and are often more or less directed by the construction of these proteins as sets of helices. If one wishes to think in this way the helices are rigid rods connecting a variety of mobile hinges and the whole can be likened to a molecular machine part. The direction in space of change of shape can be prescribed[31].

We finish this section by noting that we have not tried to describe in general terms the functional potential due to specifically placed anions on a mobile matrix in three dimensional space. It is apparent that the local potential is not exactly the same for different types of anionic groups so that we expect and find different effects of different cations through binding to polyanions of each kind. It does not appear that polyamine interaction with sugars has the same importance as with DNA and RNA.

VII. Space-Filling Folded Structures: Proteins

We have tried to show that even when we described simple cations and anions the relationship between structure and binding was very complex and that the simplification

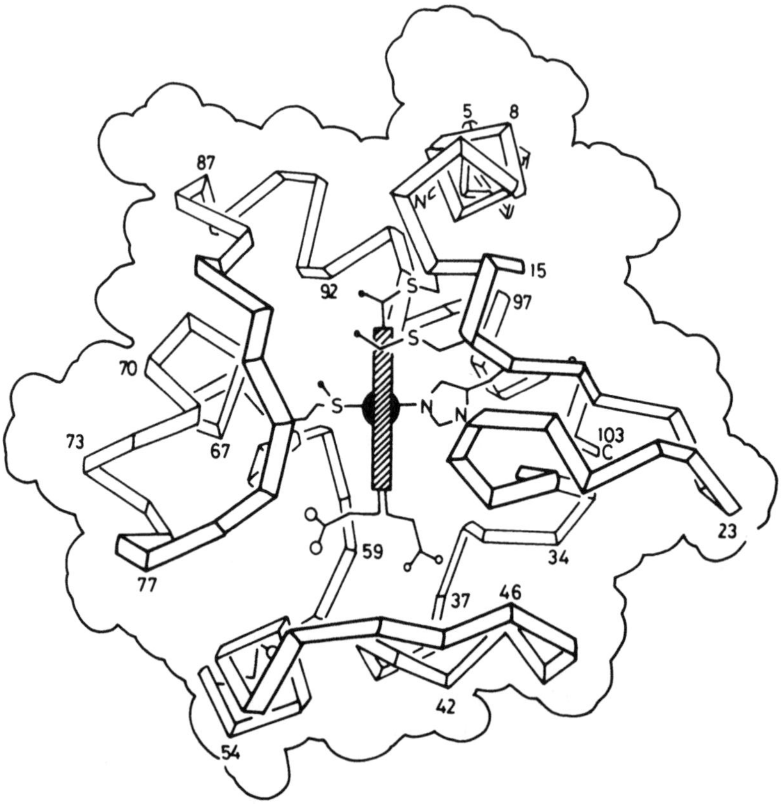

Fig. 12. The outline shape and fold of cytochrome *c*

of structure in models was always followed by a lower selectivity in the model than in observed systems. This led us to consider the selectivity of hydration of differently charged and shaped ions and then the binding of both rigid and mobile small frameworks. Although the principles are clear it remains true that quantifying them is not possible yet. When we extended the discussion to long linear polymers and planar surfaces carrying a specified type of charge, almost always negative, we again found that general theories (Manning and Gouy-Chapman) gave us a good feel for the problems but because they were devoid of structural considerations either of the simple ions, the solvent, the polymers or the surfaces they fail to generate the selective effects of biological systems. Clearly framework matching had increased in importance with the increase in size of the framework. Here the frameworks were relatively mobile and the underlying energetics hard to discern. In this next section we shall start from the different extreme of stating that there is a permanent approximately fixed position of charges on the surface of a folded space-filling molecule. This will be our initial model for a protein since in this way we can consider localised surface potential energy wells. Moreover we shall not constrain ourselves to using charges of one type but will make a patchwork of potential energy wells all over the surface where the sign of the potential is also varied. A geometric outline of such a surface is given in Fig. 12, a detailed group by group (charge by charge)

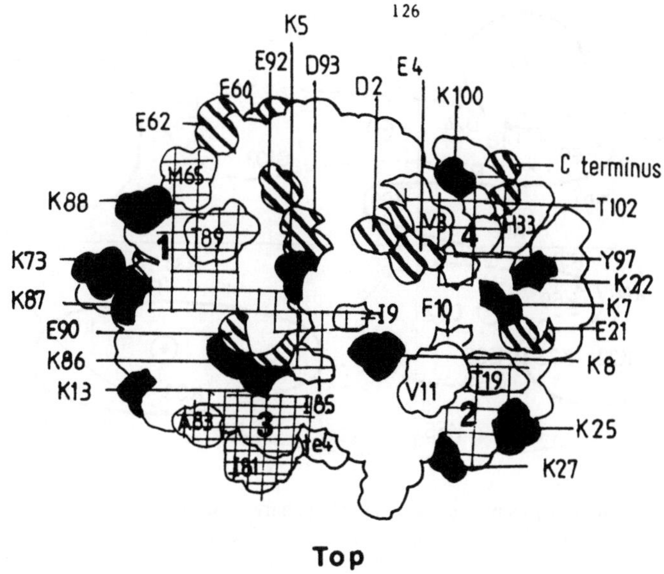

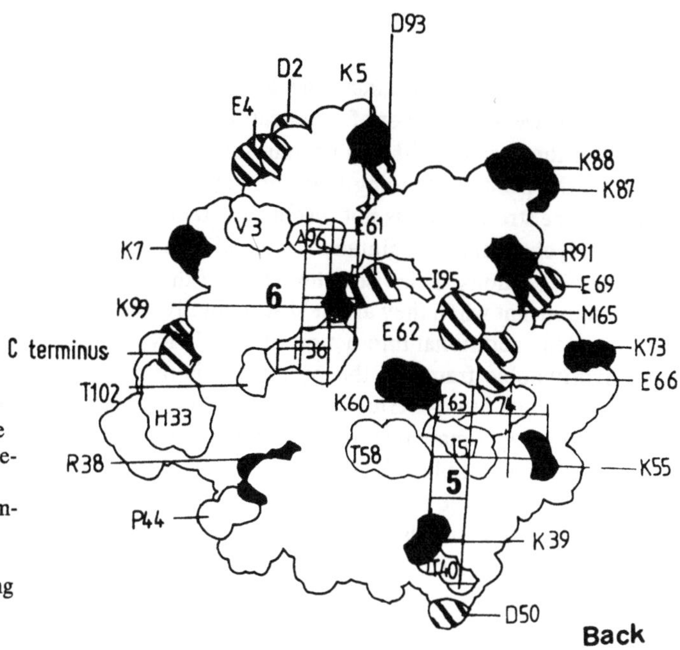

Fig. 13. Binding sites for $[Cr(CN)_6]^{3-}$ on cytochrome *c*. Filled regions are positively charged, striped regions are negatively charged. Symbols are single letter notations for amino acids. The square grid gives the binding sites, 1 to 6 for the anion

picture is given in Fig. 13 and a potential energy map is given in Fig. 14. We can give a more realistic picture of an individual protein by studying the binding of small anions and cations to proteins.

Before we do so we note that the charged groups are in fact somewhat mobile. The anions are carboxylates on short chains and the surface cationic groups are guanidinium (arginine) or ammonium (lysine). There is no tendency for such cations and anions to associate with one another when they are present in small molecules as singly charged

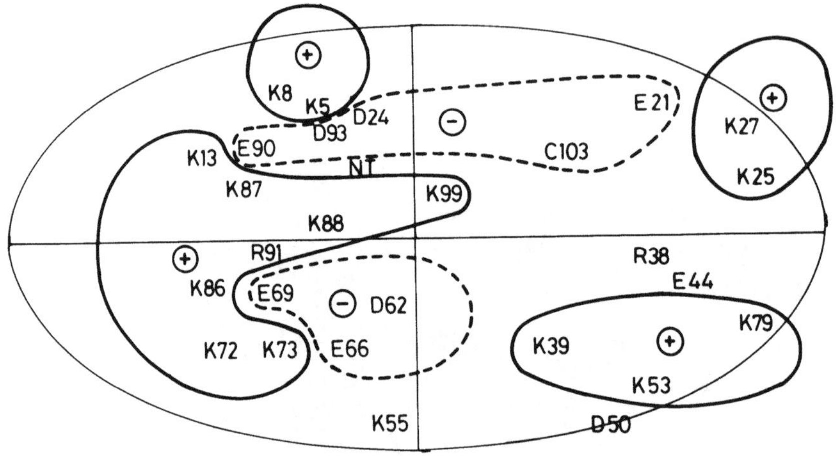

Fig. 14. A projection of cytochrome *c* surface showing charges (positive; *K* = lysine, *R* = arginine, and negative; *E* = glutamate, *D* = aspartate). The thicker lines give potential energy traps for the signs shown. They can be related to the square hatchings on Fig. 13. More detailed maps of this kind have been produced by Dr. J. Thornton and Prof. B. Atanasov

centres, Fig. 1. Association increases for diamines with dicarboxylates, Fig. 4, but this association is weak and is not of significance in biology. However, when these groups are present on the specific fold of a protein surface, the bindings are physically and chemically specific. Multiple binding between higher molecular weight species bearing these groups can be strong but it is not just electrostatic. It is noteworthy that lysine has a four-carbon CH_2 chain and that glutamic acid and arginine have two (CH_2) groups. They have hydrophobic sections. Pairwise interaction of the groups, say lysine/glutamic acid, may now be significant when they are held structurally close together on a surface of a protein because of the additional interaction of the $(CH_2)_n$ chains which can lie down upon the more hydrophobic frame of the protein. It is often found in helical structures in the peptide sequence Lys-x-x-Glu or Lys-x-x-x-Glu, e.g. in collagens. This is a trivial example of mosaic patch matching and is comparable with the diester phosphate/amine head group interactions of phosphoglycerides on the surface of membranes. A β-sheet would not appear to offer the same possibility of charge-charge interaction along a strand but could do so across strands. Glutamic acid and Lysine are more common in helices. In order to understand a surface such as that shown in Fig. 14 we must go back to the study of small ion bindings to these surfaces.

A. *Small Ion Binding to Protein Surfaces*

Tabel VIII gives the results of studies of the binding of small spherical charged cations and anions to a variety of water-soluble electron transfer proteins[32–40]. The data have been shown to be consistent with a model in which only electrostatic binding forces are operative[39, 40], as is seen from columns 4 (theory) and 5 (data) of Table VIII. Of course the proteins do not supply a fixed local charge but a somewhat localised spread out

Table VIII. Association constants of inorganic complexes with proteins (in the reduced state) at 25 °C, I = 0.1 (NaCl), pH 7.5

Protein (net charge)	Inorganic complex	K/M^{-1}	Calculated effective charge	Charge at binding site
pCu(I)[a] (−9)	[f]$[Co(III)_2]^{5+}$	16,000	3.0−	
	$[Pt(NH_3)_6]^{4+}$	22,000	4.0−	
	$[Co(NH_3)_6]^{3+}$	580	3.8−	4−
	$[Co(phen)_3]^{3+}$	167	2.8−	
	$[Cr(phen)_3]^{3+}$	176	2.8−	
[2 Fe − 2 S][b] (−17)	[f]$[Co(III)_2]^{5+}$	26,400	3.0−	
	$[Pt(NH_3)_6]^{4+}$	21,000	4.0−	
	$[Co(NH_3)_6]^{3+}$	998	4.0−	3−
	$[Cr[NH_3)_6]^{3+}$	464	3.5−	
	$[Cr(en)_3]^{3+}$	590	3.8−	
	$[Co(NH_3)_5Cl]^{2+}$	194	4.5−	
2 z4 Fe − 4 S][c] (−12)	[g]$[Pt(NH_3)_6]^{4+}$	2,500	3.4−	
	$[Co(NH_3)_6]^{3+}$	446	3.6−	ca. 3−
	$[Cr(NH_3)_6]^{3+}$	212	3.0	
cytochrome b_5[d] (−9)	[f]$[Co(III)_2]^{5+}$	16,600	3.0−	
	$[Pt(NH_3)_6]^{4+}$	14,800	4.0−	3−/4−
	$[Co(NH_3)_6]^{3+}$	600	3.8−	
	$[Cr(en)_3]^{3+}$	309	3.3−	
cytochrome *c*[e] (+9)	[h]$[Fe(CN)_6]^{3-}$	450 (I = 0.18)	~4.0+	3+/4+

[a] Parsley plastocyanin (Chapman et al., 1983(a))[33]
[b] Parsley ferredoxin (Armstrong et al., 1979)[34]
[c] *Clostridium pasteurianum* ferredoxin (Armstrong et al., 1980; 1982)[34, 35]
[d] Calves liver cytochrome b_5 (Chapman et al., 1983)[36]
[e] Horse heart ferricytochrome c (Stellwagen and Shulman, 1973; Eley, 1982)[37, 38]
[f] $[Co(III)_2]^{5+} = [(NH_3)_5Co \cdot NH_2 \cdot Co(NH_3)_5]^{5+}$
[g] pH 6.8
[h] allowance made for higher I
phen = 1,10-phenanthroline; en = ethylendiamine

charge. However in Figs. 2–4 we showed that complex ions of the sizes of those in Table VIII bind not too differently to a mobile spread out pattern of charge as to a fixed charge so long as it was not too spread out. This was not true for very simple hydrated cations and anions. The protein surfaces of Table VIII are not of high affinity for simple monoatomic cations and anions but bind well to the larger more charged anions and cations. In order to increase our knowledge of these systems we are making a detailed study using NMR of the binding of anions and cations to one protein surface that of cytochrome *c*[32, 40]. The results of some of these studies are given in Figs. 13 and 15 and in Tables IX and X. There are several sites to be considered on any one protein. The order of binding strength to the sites follows the magnitude of the charge on the small molecule and the local charge of the protein and is largely independent of the charge-bearing group. Thus for this series of small molecules it is the electrostatic term which is dominant. We know however that chaotropic agents also bind to cytochrome *c* and we are now

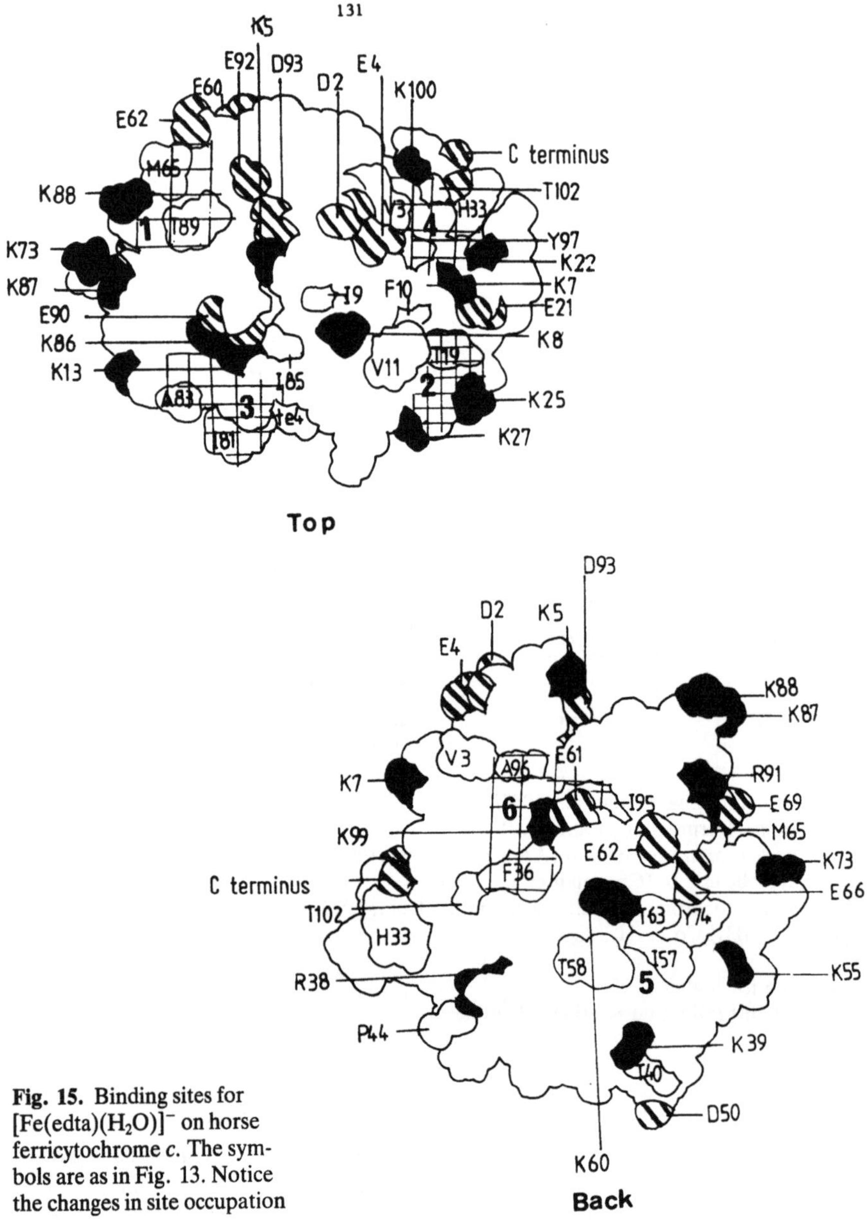

Fig. 15. Binding sites for $[Fe(edta)(H_2O)]^-$ on horse ferricytochrome *c*. The symbols are as in Fig. 13. Notice the changes in site occupation

extending our work to a study of small probe molecules containing both hydrophobic and charged zones e.g. ATP. We expect to observe specific effects and that the general theories using electrostatics will fail for the same reasons as given above.

We can consider the experiment of the binding of $[M^{III}(CN)_6]^{3-}$ to the many sites on cytochrome *c* as an inspection of the positive electrical potential energy of the protein surface, see Fig. 14. Binding sites can be viewed as minima in this surface energy[32, 40]. Put another way we can imagine that the $[M^{III}(CN)_6]^{3-}$ is a ball of radius 5 Å which is

Table IX. Residues[a] in the binding sites for Mn^{2+}

Sites	I	II	III	IV
Probable charged residues involved in the site (from crystal structure)	E104, E61	D2, E4 E92 (E90, D93)	E66, E62 E69, (E61)	E21
Other residues involved in the site for which broadenings have been observed	I95, T102 A101, F36	Nac, T89 V3/V20, A96, I9	M65, Y74 I95, F36	T19, (Y97) (A101), (V20)
Residues near to the site for which broadenings have not been observed	T63 L35	V11	T58	A15

[a] Only surface residues are considered. E = Glutamate, D = Aspartate. The other letters are single letter codes for uncharged amino-acids. The numbers refer to position in the sequence.

Table X. Residues[a] in the binding sites for anions[40)]

Sites	1	2	3	4	5	6
Probable charged residues involved in the site (from crystal structure)	K86, K87 K88	K25, K27	K13, K72 K79	K5, K7 K8	K55, K39	K99, K100
Other residues involved in the site for which broadenings have been observed	I9[b], I85 M65	T19, A15 H26, T28	I82, F82 A83, I85 T28, T47[b] (T78)	Y97[b], V11 Nac, T89[b]	Y74, I75[d] T40[d]	F36, I95[c] (A96)
Residues near to the site for which broadenings have not been observed		A43	A51	F10	I57, T58	T102

[a] Only surface residues are considered
[b] Not observed in $[Fe(edta)(H_2O)]^-$ binding
[c] Not observed in $[Gd(dpc)_3]^{3-}$ binding
[d] Not observed in $[Cr(CN)_6]^{3-}$ binding
K = lysine. See Table IX for the other codes.

rolled all over the surface of the protein and which is able to read out the positive potential energy contours to provide a map. We can generate various maps of this kind, rolling bigger balls such as $[Gd(dipicolinate)_3]^{3-}$ or balls of opposite charge e.g. $[Cr(NH_3)_6]^{3+}$. The gadolinium complex finds the more hydrophobic positively charged regions while the chromium hexamine draws out the negative potential energy map. We can then take ions with shapes such as $[Fe(EDTA)]^-$ and roll them, assuming them to be more like small cylinders, over the surface of cytochrome *c*. They will find regions which have dipoles matching that of $[Fe^{3+}(EDTA^{4-})]^-$. In fact there is a great variety of such maps which do not relate to any one fixed surface but a combination of structural properties. The surface is partly adjustable and interactive with the inspection device. It could be asked, "Is such a surface in any way selective?" The answer is very much so, as we show later, but it requires a large complicated reciprocal surface to exhibit the full selectivity.

There is a further feature of such side-chains as those of lysine and to a lesser degree arginine and glutamate which we mentioned before. Their aliphatic chains are very mobile. The mobility of the four (CH_2) groups of lysine make its headgroup, ammonium centre, able to move over a wide area. This mobility may be a disadvantage if selectivity of binding for especially a small molecule is desired but it does allow fast reaction. The group mobility becomes less detracting from selectivity of binding the larger the area of surface which is matched with a partner, see Figs. 1 to 5. It is here that the idea of dynamic mosaic-patch matching is important since there can be considerable selectivity with very fast reaction. The selectivity is especially high if the local dynamic patches are of very different physical property e.g. if alternating hydrophobic and hydrophilic groups are present. Viewed in this light it is possible to conceive how proteins can be organised in water or can partition into lipid bilayers and yet become organised in an organisation which may be rapidly rearranged as in mitochondrial or thylakoid membranes. The relative movement of protein surfaces as well as the selectivity of binding must be an essential feature of the proteins involved in the machinery of the cell.

Another important consequence of the mobility of groups is that it can allow surface diffusion of other molecules to occur easily between potential energy wells with considerable selectivity. Surface diffusion is important for proteins such as cytochrome *c* on the surface of the mitochondrial membrane but it can also be important for substrates entering or leaving active sites. We are examining this possibility at present using NMR methods.

B. Protein-Protein Interactions

In the course of the above passage it should have become clear that we do not regard the surface/surface interactions of proteins as anything but highly selective, i.e. to a degree which can not be revealed by electrostatic treatments. If we start an analysis of selectivity of a protein surface from its ability to bind to a very small ion with high selectivity then rather little mobility can be permitted in the cavity site, e.g. calmodulin[31)]. Increasing the size of the substrate to a multidentate molecule or a large ion allows selectivity to be achieved even when the mutual relationship of protein and small molecule are not so well prescribed by rigid geometries e.g. acetylcholine is not at all rigid yet its binding to its receptor is very selective. We presume that such selectivity still demands a cavity. As the substrate increases in size and mobility, for example a substrate protein, the selectivity can be increased by increasing the number of points of attachment but clearly the mobility of the binding groups includes that of both proteins, one acting as the substrate. Only when the on/off reactions must be very selective – almost specific, and can be slow, is precisely defined structure necessary before connection is made. This is the case of a dye fitting a mould as perhaps in the matching of protein inhibitors to proteases[41)]. Again we visualise this in terms of a large cavity. The more general case of protein matching is more akin to the fitting of a hand into glove – dynamic mosaic patch matching[24, 33)]. Different proteins must be able to fit very selectively to a wide variety of different surfaces which include other proteins, membranes, DNA and RNA and yet each polymer must move about rapidly and the concept of a cavity is itself no longer applicable. To repeat it is the use of multiple point matching of mobile centres which enables the surface to be free from competition from small molecule interactions and still maintain selectivity and mobility. Such protein/protein surface matching is not competitively inhibited by the presence of millimolar divalent ions such as Mg^{2+}, Ca^{2+} and HPO_4^{2-} which bind with $K \leqslant 10^2$ while protein/protein interactions can exceed 10^6. However, extreme specificity even in protein/protein interaction will need closer matching of surfaces, i.e. the use of cavities, reducing mobility in the interaction as in antigenic-protein antibody fitting[42)]. Protein/protein matching in transducing organisations needs recognition and mobility but this is not so true of the organisation of inhibitor proteins on receptors.

Various groups have attempted to provide diagrams of protein surface-surface interaction using computer modelling[43–45)]. These models usually assume very simple electrostatic charge matching without regard for other forces between the frameworks. This is an extremely dangerous procedure since many proteins are known to bind together through the most hydrophobic parts of their surfaces. The problem arises from the fact that there are no energetic considerations going along with the matching process. We shall illustrate this elsewhere by an analysis of the packing of proteins in crystals. The electrostatic interactions are but one term in the mosaic of interactions. We have shown above how this point can be approached experimentally using small complex ions. A large part of the difficulties arise from the use of a bulk solvent model, dielectric D, when only a molecular treatment of the solvent is adequate.

C. Electrostatics Through Proteins

A large number of investigators have tried to analyse the interaction energy of protein charges which could not be said to lie within the water solvent, that is they are not fully exposed residues. Notable attempts at different aspects have been made by Berensen[46)] (especially charges on the ends of helices), Kirkwood and Tanford[47)] (partially buried charges), Gurd[48)] (ionisation of histidines) and several theoreticians[49–51)]. The central problem in any such quantitative analysis is the description of the medium inside the protein. Above it has been assumed that the interaction energy of anions with lysines on the surface of a protein can be treated in the same way as ion-pairs in water i.e. a dielectric constant, D, of 80 has been assumed as in Debye-Hückel, Bjerrum, Manning and Gouy-Chapman analyses. This is clearly inapplicable once the charge is partially buried. Gurd and his coworkers especially have used the Kirkwood-Tanford approach with $D = 5$ inside the protein and $D = 80$ outside the protein to study partially buried histidine residue pK_a values and the effect of changes in charge on the surface of the protein upon these values. The quantitative agreement is remarkable. Unfortunately the nature of the discussion has been questioned[47)] and alternative approaches using more fundamental statistical mechanical treatments are being attempted. The difficulty here lies in the complexity of the problem. It seems that the treatment where the fractional exposure of a group to water is found by rolling a ball of the radius of the water molecule and so finding the ratio, η, of the area of contact relative to the total area of the surface is as good a starting point as any[48)]. The medium effects can then be treated by letting a fraction of the group $(1 - \eta)$ be in the organic solvent of the protein core. This treatment will not affect the description of *surface* lysines, arginines and most carboxylates which are far from the hydrophobic core but it is very important if these groups lie very close to hydrophobic regions. In cytochrome *c* for example all the charged groups except two or three are exposed on the surface. The two propionic acid groups of the haem however are buried and one of them at least is likely to be charged and placed opposite a buried arginine which carries the counter charge. Moore[52)] has discussed the importance of this centre in the redox reactions of cytochrome *c* using the notion of a low dielectric constant. There are other buried charge-relays in proteins such as the serine proteases. In this article we shall not do more than draw attention to their existence but we note that the discussion of the problem in quantitative terms is full of problems due to the heterogenous nature of proteins.

D. *Protein Helix Dipoles and Protein Dipoles*[46, 53)]

Another problem which concerns the interior of proteins more than the exterior is the existence of large dipoles due to the summation of the ordered dipoles of helical stretches of amino acids. The use of these helix dipoles in the binding of small ions in the interior of proteins has been analysed especially by Berendsen and his coworkers[46)]. At the present time their significance has not been fully evaluated but it seems unlikely that there will be a contribution to the binding of molecules on the surfaces of proteins which is our major concern. Once again in the interior of proteins the understanding of electrostatic energies is bedevilled by the description of the medium.

A very different problem arises as to when it is appropriate to sum the whole network of charges on a protein surface so as to give a dipolar field around it. The asymmetry of charges on some proteins e.g. cytochrome *c* is such that a very large dipole is generate with an axis which has a particular relationship to the active site[53)]. The dipole will generate long-range organisational energies and is possibly best used in the discussion of general salt effects and the approach to molecules to one another. The local surface charge distribution remains the correct reference for the electrostatics of binding a surface and an ion in an ion-pair.

E. *Proteins Carrying Phosphate*

This is the last anionic group to be found on the surface of biopolymers. The phosphate group is a terminal anion but is very different from sulphate. The affinity for all simple cations is high e.g. H^+, Mg^{2+} and Ca^{2+}. Given the presence nearby of other phosphates, carboxylates or even –OH groups these cations can bind and screen the phosphate group. In the absence of such extra centres the single phosphate group can bind quite strongly to amine centres. The balance between simple cation and amine binding is much more even than for sulphates and phosphate esters and is critically dependent upon the organic moiety which carries the phosphate and the amine. It is this fact that makes phosphorylation and dephosphorylation such a fine control of biological organisation. By introducing phosphate into polymer/polymer systems which normally carry only diester phosphate or sulphate (or even carboxylate) biological systems increase the electrostatic components of binding (high local charge) at the expense of the hydrophobic component. Thus we find highly phosphorylated proteins in association with teeth, bone and milk (Ca^{2+} ions) outside cells and inside cells we find Mg^{2+} ATP. However where proteins generate positively charged cavities we expect to find special fitting due to phosphate (much as observed for calcium) and this will provide a highly selective interaction. We must always remember that phosphorylation gives a time dependence too to electrostatics since it is under enzymic control.

F. Folded Proteins, Histones, and Linear Polyelectrolytes, DNA

The interaction of histones and DNA cannot be highly specific since the same combination of histones binds equally to all DNA sequences. Moreover interaction with the nucleotide bases is limited by the building of the double helix. NMR evidence as to the nature of the isolated histones shows them to be folded largely in helices but maintaining a highly positively charge surface and tails. The tails are mobile.

The expression of DNA depends upon the separation of the histones from the DNA itself. It is known that changes in salt concentrations bring about their separation and in keeping with the notion that the histone: DNA interaction is largely electrostatic it is high salt which causes separation. In vivo expression of DNA is not managed in this way – there is no change of salt concentration during cell expression or cell division. We must seek for alternative ways of controlling the electrostatic forces. One method is to introduce by synthesis competing proteins or by conformational change.

There are however more direct ways of interfering with the delicately maintained electrostatic interactions between DNA and histones which require changes in the charges on the histone. There is the well-known example of histone phosphorylation. The reaction takes place in the tail of the histone changing its positive charge by two units locally. We can well imagine that the tail is then dissociated from the DNA (negatively charged DNA will repel the phosphate). An alternative way in which to change charge and to introduce hydrophobic selectivity is by acetylation of the amino groups.

We can compare this neutralisation of positive charge with the neutralisation of negative charge by calcium. The effect in the latter case was transmitted to the whole of the contractile system through the motions of the helices of the troponins or through those of calmodulin to kinases[31)]. The histones resemble the calmodulins and troponins C in that they too are helical proteins. A change of charge in the terminal region preventing interaction with the negative charge of DNA could readily reduce the cooperative DNA/histone interaction allowing the DNA to be more accessible and more easily unwound and expressed. We shall expand on this view of histone function in a later article.

A very interesting case of the combined use of Ca^{2+} and basic groups of proteins in the stabilisation of a packed form of a polynucleotide viral RNA is in tomato bushy stunt virus. Here the Ca^{2+} ions interact with carboxylate groups of the protein while the amino side-chains of the protein bind to the diester phosphate groups of the RNA.

VIII. Cell/Cell Interaction

The idea of dynamic mosaic-patch matching is general to the selectivity of interaction of large surface areas. It is less valuable as a concept for small substrate specificity on a surface where the best matching will arise from cavities of more or less rigid construction. As we stressed above however such rigid systems have the striking disadvantage of being slow. Cell/cell interaction provides the possibility of huge surfaces over which many parts of a dynamic mosaic may interact. One example is given by the sponges[29)] where calcium no doubt plays a role as an electrostatic cross-linking agent. Here too we can re-introduce the polysaccharides which also have a selectivity due to their partial folding and so can take part in dynamic mosaic charge matching. So far the theory of cell-cell interaction is so poor that there is no point in elaborating here upon the obvious extension from protein/protein interaction to that between cells.

IX. Summary: The Biological Problem of Electrostatic Control

It is clearly the case that a large number of control systems in biology are based at least in part upon the selectivity of electrostatic forces. The selectivity arises from charge and size considerations in the first place and we must take into account hydration, i.e. the size of water. The severity of the problem can be seen by looking at Table VI which shows the number of anions and cations of different kinds which interact with one another. Many of these interactions occur at the same time. There are some highly selective associations – how do they arise? We must observe firstly that the same anions and cations are not to be found in all parts of space, Eq. (3). By separating cations amongst themselves and anions amongst themselves and by regulating the absolute levels of ions individually the probability of some associations in compartments is increased and that of some others greatly diminished. Table XI sets out the cation and anion distributions which are known to be important generally and Table XII sets out some locally important features of particular organisms or organs. As a consequence of these separations in biology the following general situation exists: There is little or no association between any of the simple small biological anions and any of the simple small biological cations (with the exception of ATP, $P_2O_7^{4-}$ and so on with Mg^{2+}). Electrostatic forces per se for these ions are completeley screend out by water. This means that organisation (of charges) in polymers takes on a special significance since organisation (of charges) on a framework can still generate powerful electrostatic binding and binding selectivity. The first level of organisation is a single anionic or cationic charge placed in a designed almost rigid cavity e.g. by the fold of a protein. This gives a radius-ratio selectivity and is used to capture simple cations or anions highly selectivity and with high affinity e.g. CaIBP, Table XIII. Many receptors may well be similar. We can increase the number of charged centres in the organisation of the receptor from say one to six without altering the selectivity or binding principles for a simple single cation e.g. the calmodulin hands select for calcium and against magnesium. The affinity and the selectivity are under the control of the cavity forces; only a part of the control is from charge-charge interaction, compare calcium binding to phospholipase A.2. with CaIBP, Fig. 16.

Table XI. Spatial constraints upon ion distributions

Outside Cells	Inside Cells
Na^+, Ca^{2+}	K^+, Mg^{2+}, polyamines
Cl^-	Phosphate anions (esters etc.)
Sulphated Sugars	DNA, RNA diester phosphates
Carboxylated Polysaccharides	Highly Carboxylated Proteins
Aminoglucosoglycans	Histones
Basic myelin protein	

N.B. Phospholipids are relatively evenly spread in all membranes and both inside and outside cells.

Table XII. Selective biopolymer interactions

Anion	Cation	System
DNA (P)	Histones 2/3/4	Nucleosomes
	Histone 1	Between Nucleosomes
	Polyamines	Sperm (Single Strand)
RNA (P)	Set of Basic Proteins	Ribosome
Myelin Membrane (C)(S)	Basic Protein	Myelin Sheath
Polysaccharides (C)(S)	Calcium	Contact Surfaces
Viral Coat Protein (C)	Calcium	Virus Package

C = carboxylate, P = phosphate diester, S = sulphate
Note we have treated the proteins such as histones and basic proteins as mobile polymers, which is only true in part.

Table XIII. Outline of factors affecting charge interactions

Basic electrostatics	Hydration	Cavity	Spatial charge distribution
z_1z_2 dr_{12} Bulk dielectric	Radius ratio effects Cation hydration Molecular solvent	Organic molecule hydration Restricted entropy	Shape of multielectrolyte Flexible molecules (Fold)
Debye-Hückel Bjerrum Gouy Chapman Manning	Pauling Williams	Williams Lehn Cram	This paper

N.B. No account is taken here of forces other than those due to electrostatic fields.

If the framework has a less rigid structure then site selectivity for small cations will begin to be lost although strong binding can still be generated provided that there are several charges e.g. for EDTA, prothrombin tail, and ATP Mg $\geqslant$ Ca. The group of charges must be relatively close together. Selectivity can be regained only by introducing non-charged interactions, for example, neutral co-ordination centres, e.g. in EGTA and

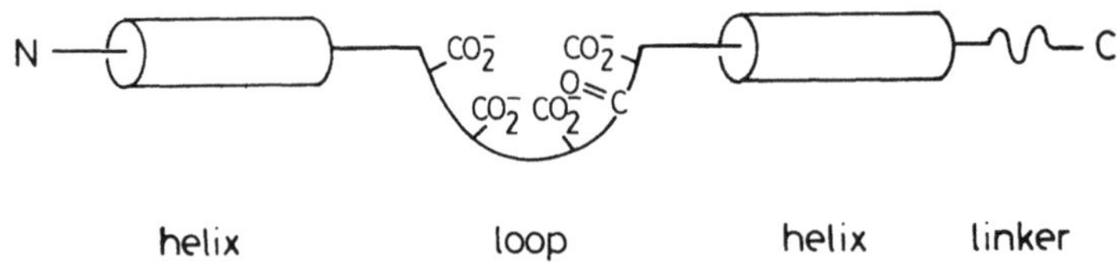

Fig. 16. The characteristic helix-loop-helix-linker segmental structure of same calcium trigger proteins. The loop is the highly selective calcium input site, the helices are rods for information transmission and the output linkers connect to other proteins through dynamic mosaic patch matching with many electrostatic centres as well as hydrophobic regions

calmodulin, Ca Mg. However, if organic anions and cations are now considered then they begin to bind appreciably and selectively Figs. 2–4.

Next consider the opposite extreme of a rigid rod of equally spaced anionic charges. There is no selectivity, no cavities are possible and simple amines RNH_3^+ interact as poorly as Na^+ and K^+. The interaction can be treated by normal Debye-Hückel treatments when the anions are widely dispersed on the rod. The competition from Mg^{2+} or Ca^{2+} is not severe since $[M^+] > [M^{2+}]$ by one hundred-fold and all binding is weak. The binding becomes interesting (Manning theory) only as the anions come together to shorter distances than about 10 Å separation when the cations are predicted to *condense* around the rigid rod anions. At larger distances of separation the negative charge of the rigid rod behaves non-selectively possessing the Debye-Hückel atmosphere of positive charges (i.e. the same as the ionic atmosphere for an isolated charge). However a new selectivity can now arise. The binding of a rigid rod of positive charges which had exactly the same charge separation as a negatively charged rod could now be very strong so that selectivity of two organic rigid rod matrices of opposite sign and equal spacing is a source of high selectivity against simple cations. This is charge matching, see section, which we can not treat quantitatively. We now have different grounds for selectivity on a rigid framework

(a) Very high local charge: M^{2+} not polyamines or M^+ bind.
(b) Special local cavities of low charge can bind any single ion M^+, M^{2+} or RNH_3^+ with high selectivity due to cavity design.
(c) Organised surfaces of highly separated charges can only bind other surfaces of equally spaced charges and will not be screened by simple cations.
(d) Organised surfaces of closely packed charges will CONDENSE with their screening charges which can be simple ions or correctly organised charge systems of opposite sign.

(Note under (c) and (d) we shall need to consider hydrophobic regions of the polymers later.) These rules follow from models we can build but are only partly in accord with theoretical expectation, see above. We next turn to non-rigid frames.

We have seen the effect that as the anions are spaced further apart on a rigid chain then the affinity for simple inorganic cations falls but the affinity for polyamines on a rigid frame may increase Figs. 2–5. This arose through charge matching. However charge matching did not in fact need rigid frames, Figs. 2–5. There could be a large number of ways in which two mobile frames could match. Be this as it may charge matching by exactly equal positioning of charges on mobile frameworks is also likely to give the greatest strength of binding. In other words the degree of flexibility, the number of charges involved, their average spacing on both the interacting species will decide the strength and the selectivity of binding. We suspect that where it is possible to set the spacing, flexibility, and charge numbers so that competition between the chain or protein organic amines and the simple cations for chain or protein anions is in critical balance then a control system can be developed. One form of control tests in the controlled movement of cations across membranes. Although it may be useful to set up this type of control in models using systems of well-defined structure it is not necessary to work within any such limitations and highly mobile chains can give sufficient thermodynamic selectivity as we have shown. The same considerations apply to two dimensional surfaces carrying charge of one kind only. Finally we turned to mosaics of different charge type and showed the principles of dynamic mosaic charge matching which gave high binding strength and selectivity between large areas of surfaces.

We saw in the earlier sections how the nature of radius ratio effects in crystal lattices led us to conceive the idea of cavities in molecule frameworks as sources of high selectivity. The selectivity of the formation of a crystal was associated additionally with a phase change which was due to cooperative forces[54]. We can now proceed in the opposite way from a consideration of charge matching selectivity of pairs of frameworks to huge condensed organised systems notching that selectivity enhancement similar to that in crystals will arise if the interactions become cooperative. There will not be a repeating lattice however. We have shown that pair-wise open-faced mosaics of opposite character, charge etc., can be made to match and are selective. We can now imagine identically sized cubes, six open faces but each face different, binding to other cubes with similar dimensions and faces of complimentary character packed into pairs (one to one linear association). We can extend the packing to higher levels of organisation, three dimensional, when it is clear that the pair-wise packing of three cubes A, B and C by shared faces can be assisted cooperatively by binding D to complete a square of cubes. Cooperatively in a lattice does not depend on symmetry of packing, nor does it depend on identical shapes of objects. Well packed objects of irregular features are possible. The selectivity of the whole assembly can exceed the sum of the pair-wise interactions. At present it would not seem to be possible to formulate rules to govern the shape or form of the objects which can be generated. The ideas are common to all packings from proteins to cells and the general sense is that extreme selectivity here can be a property of complexity of surfaces. In fact there is built up by the non-identical units, proteins, a phase which has boundaries. The peculiarities of these phases are not yet understood whether we have to describe parts of cells or whole organs. It remains true that electrostatics plays a large part in their stabilisation[54] but until we can handle the selectivities which arise from the size and shape of molecules, especially water, the models we use will always hide the specificity of biological interactions.

X. References

1. Debye, P., Hückel, E.: Physik. Z. *24,* 185 and 305 (1923)
2. Onsager, L.: Trans. Faraday Soc. *23,* 341 (1927)
3. Davies, C. W.: Ion Association, Butterworths London (1962)
4. Bjerrum, K.: Danske Vidensk Selsk Mat. fys Medd. *9,* 7 (1926)

5. Williams, R. J. P.: J. Chem. Soc. (London), 3770 (1952)
6. Williams, R. J. P.: Quart. Revs. Chem. Soc. (London) *24,* 331 (1970)
7. Tam, S.-C., Williams, R. J. P.: J. Chem. Soc. Faraday Trans. I, *80,* 2255 (1984), and Tam, S. C.: D. Phil. Thesis, Oxford 1984
8. Phillips, C. S. G., Williams, R. J. P.: Inorganic Chemistry, Vol. I, Chap. 20., Oxford Univ. Press 1966
9. Pauling, L.: The Nature of the Chemical Bond. Cornell Univ. Press, Ithaca 1938
10. Williams, R. J. P.: Structure and Bonding *50,* 77 (1982)
11. Williams, R. J. P.: The Analyst *78,* 586 (1953)
12. Pedersen, C. J., Frendsdorff, D.: Angew. Chem. *84,* 16 (1972)
13. Behr, J.-P., Lehn, J.-M., Vierling, P.: Helv. Chim. Acta *65,* 1853 (1982) and refs. therein
14. Cram, D. J., Cram, J. M.: Accounts Chem. Res. *11,* 8 (1978)
15. Cullinare, J., Gelb, R. I., Margalis, T. N., Zompa, J. L.: J. Amer. Chem. Soc. *104,* 3048 (1982)
16. Breslow, R., Rajagopalen, R., Schwarz, J.: ibid. *103,* 392 (1981)
17. Kimura, E., Sakonaka, A., Yatsumami, T., Kodama, M.: ibid. *103,* 3041 (1981)
18. Kumura, E., Sakonaka, A., Kodama, M.: ibid. *104,* 4984 (1982)
19. Manning, G. S.: J. Chem. Phys. *51,* 924 (1969), and Manning, G. S.: Accounts Chem. Res. *12,* 443 (1979)
20. Delville, A., Gilboa, H., Laslo, P.: J. Chem. Phys. *77,* 2044 (1982)
21. Delville, A., Laslo, P.: Biophys. Chem. *17,* 119 (1983)
22. Bräulin, W. F., Nordenskidd, L.: Europ. J. Biochem. (1985) to be published
23. Molbrook, S. R., Sussman, J. L., Warrant, R. W., Church, G. M., Kim, S. H.: Nucleic Acid Res. *4,* 2811 (1977)
24. Schweizer, M. P., De, M., Pulsipher, M., Brown, M., Reddy, P. R., Petrie, C. R., Chheda, G. B.: Biochim. Biophys. Acta *802,* 352 (1984)
25. Ramesh, N., Brahmachari, S. K.: Febs Letter *164,* 33 (1983)
26. Odijk, T.: Macromolecules *13,* 1542 (1980)
27. Sposito, G.: The Surface Chemistry of Soils, Oxford Univ. Press 1984, Chapter 5
28. McLaughlin, K.: Current Topics in Membranes and Transport *9,* 71 (1977)
29. Williams, R. J. P.: in Calcium-Binding Proteins (B. de Bernadi et al., eds.), Elsevier Science Pub. Amsterdam 1983, p. 319
30. Egmond, M. R., Williams, R. J. P., Welsh, E. J., Rees, D. A.: Europ. J. Biochem. *97,* 73 (1979)
31. Dalgarno, D., Klevit, R. E., Levine, B. A., Williams, R. J. P.: Trends in Pharmacological Sciences *5,* 266 (1984)
32. Moore, G. R., Williams, G., Williams, R. J. P.: Frontiers in Bio-organic Chemistry and Molecular Biology, ed., Y. Ovchinnikov, Elsevier, p. 31, 1985
33. Chapman, S. K., Watson, A. D., Sykes, A. G.: J. Chem. Soc. Dalton Trans. 2543 (1983)
34. Armstrong, F. A., Henderson, R. A., Sykes, A. G.: J. Amer. Chem. Soc. *101,* 6912 (1979)
35. Armstrong, F. A., Henderson, R. A., Ong, H. W. K., Sykes, A. G.: Biochim. Biophys. Acta *681,* 161 (1982)
36. Chapman, S. K., Davies, D. M., Vuik, C. P. J., Sykes, A. G.: J. Chem. Soc. Chem. Comm. (London) 1983, 868
37. Stellwagen, E., Shulman, R. G.: J. Molec. Biol. *80,* 559 (1973)
38. Eley, C. G. S.: D. Phil. Thesis, Oxford Univ. 1982
39. Chapman, S. K., Sinclair-Day, J. D., Sykes, A. G., Tam, S.-C., Williams, R. J. P.: Chem. Comm. Chem. Soc. London, 1152 (1983)
40. Tam. S.-C.: D. Phil. Thesis, Oxford Univ., England 1984
41. Tsetlin, V. I., Karlsson, E., Utkin, Y. N., Pluzhnikov, K. A., Arsenier, A. S., Surin, A. M., Kondakov, V. V., Bystrov, V. F., Ivanov, V. T., Ovchiniskov, Y. A.: Toxicon *20,* 83 (1982)
42. Moore, G. R., Williams, R. J. P.: Trends in Biological Sciences *10,* 96 (1985)
43. Matthew, J. B., Weber, P. C., Salemma, F. R., Richards, F. M.: Nature *301,* 169 (1983)
44. Salemma, F. R.: Ann. Rev. Biochem. *46,* 299 (1977)
45. Zielenkiewicz, P., Rabczenko, A.: J. Theoret. Biol. *111,* 17 (1984)
46. Berendson, H. J. C.: Abstracts of the Symp. on Electrostatic Effects in Proteins 8th IUPAB Congress, Bristol 1984
47. Tanford, C. and Kirkwood, J. G.: J. Amer. Chem. Soc. *79,* 5333 (1953)

48. Matthew, J. D., Hanania, G. H. I., Gurd, F. R. N.: Biochemistry *18,* 1919 (1979)
49. Warshel, A., Russell, S. T., Chung, A. K.: Proc. Natl. Acad. Sci. *81*, 785 (1984)
50. Thornton, J. M.: Nature *295,* 13 (1982) and Barlow, D. J., Thornton, J. M.: J. Molec. Biol. *168,* 867 (1983)
51. Rogers, N. K., Sternberg, M. J. E.: Molec. Biol. *174,* 527 (1984)
52. Moore, G. R.: Febs Letters 161, 171 (1983)
53. Koppenol, W. H., Margoliash, E.: J. Biol. Chem. *257,* 4426 (1982)
54. Williams, R. J. P.: Biochem. Biophys. Acta *416,* 237–283 (1975)

Chemistry of Tetra- and Pentavalent Chromium

Kamalaksha Nag and Satyendra Nath Bose

Department of Inorganic Chemistry, Indian Association for the Cultivation of Science, Calcutta 700032, India

This review provides information on synthesis, reactivities, solid state properties, and electronic and molecular structure of chromium(IV) and chromium(V) compounds that have been isolated in pure form. Spectroscopic properties, especially EPR of Cr^{5+} and Cr^{4+} ions doped in various host lattices, are also discussed. However, compounds that generated in solution as transient species, or occured as intermediates in redox reactions, lie outside the scope of discussion. The types of compounds which received major consideration are: oxo-compounds, peroxo complexes, halo and oxyhalo complexes, alkoxides, amides, alkyls, hydroxy carboxylates and macrocyclic compounds. Literature through 1984 is covered. It will be apparent from this article that certain aspects of chromium(IV) and chromium(V) chemistry will get enriched in future.

1 Introduction

Although the known oxidation states of chromium range from -2 to $+6$[1], the most prevalent states are $+6$, $+3$ and $+2$. Chromium compounds with formally -2 and -1 valence states are extremely rare, however, with π-acid ligands a fairly good number of compounds are known for 0 and $+1$ states[1,2]. The oxidation states $+4$ and $+5$ are generally held uncommon, yet their chemistry have grown substantially in the last 20 years to merit a detailed consideration. Chromium in all of its valence states does not show much resemblance with its Group VI congeners, molybdenum and tungsten. In general, the lower oxidation states are more stable for chromium, whereas the higher oxidation states for molybdenum and tungstane are more common and much less susceptible to oxidation. Chromium(IV) and chromium(V) compounds are generally sensitive to air and moisture, and therefore usually require special handling procedure. The coordination number of chromium in these two oxidation states span from 4 to 8. The usual steric environments for chromium(IV) compounds are tetrahedral and to a lesser extent octahedral, and for chromium(V) compounds tetrahedral or square pyramidal. The stereochemistries for a few representative types of compounds are given in Table 1.

Cr^{4+} and Cr^{5+} ions being hard acids show strong preference for oxygen, fluorine, chlorine and ligands containing nitrogen. A large number of tetraalkyl compounds for

Table 1. Coordination number and stereochemistry of chromium(IV) and chromium(V) compounds

Oxidation state	Coordination number	Stereochemistry	Examples
+4	4	Tetrahedral	Sr_2CrO_4, $Cr(OBu^t)_4$, $Cr(CPh{=}CMe_2)_4$
	5	Square pyramidal	CrO(TPP)[a]
	6	Octahedral	$BaCrO_3$-polytypes, $Sr[CrF_6]$, $Rb_3[CrF_7]$
	7	Pentagonal bipyramidal	$[Cr(O_2)_2(NH_3)_3]$, $K_3[Cr(O_2)_2(CN)_3]$
	8	Dodecahedral	$[CrH_4(dmpe)_2]$[b]
+5	4	Tetrahedral	$M^I_3CrO_4$, $Ca_5(CrO_4)_3X$ (X = F, Cl, OH) Ca_2CrO_4Cl
	5	Square pyramidal	$[Ph_4As](CrOCl_4)$, CrN(TPP)[a], $[CrO(salen)][PF_6]$[c]
	5	Intermediate between square pyramidal and trigonal bipyramidal	$K[OCr(O_2CCOMeEt_2)]$
	6	Octahedral	$K_2[CrOCl_5]$, $[CrF_5]_\infty$
	7	–	–
	8	Dodecahedral	$K_3[Cr(O_2)_4]$

[a] H_2TPP = Tetraphenyl porphyrin; [b] dmpe = Dimethylphosphinoethane;
[c] H_2salen = Bissalicylaldehydeethylenediamine

chromium(IV) also exist. Compounds with soft donors such as sulfur, phosphorous and arsenic are virtually nonexistent, except for a few unusual chromium(IV) compounds obtained with a phosphorous and some sulfur containing ligands.

Even through development of the chemistry of chromium(IV) and chromium(V) compounds took place mainly in the past two decades, yet discovery of some of the most important compounds were made long before. For example, a chromium(IV) peroxo complex, $[Cr(O_2)_2(NH_3)_3]$ was prepared by Wiede in 1897[3]. In 1905 Reisenfeld and coworkers[4] reported the formation of a chromium(V) peroxo complex, $K_3[Cr(O_2)_4]$. In the same year Weinland and coworkers[5] prepared salts of the oxychloro anion, $[CrOCl_5]^{2-}$. Interestingly, as early as in 1859 Wöhler[6] identified a ferromagnetic chromium oxide, that, however, was not characterized as CrO_2 until 1935 by Michel and Bernard[7]. The preparation of ternary oxides of chromium(IV) and chromium(V) with alkali and alkaline earths by Klemm[8a] and Scholder[8b] in the early fifties also mark an earlier stage of development.

The intent of this review is to collect together and analyze information on chromium(IV) and chromium(V) compounds that have been isolated in pure form. Synthesis, reactivities, solid state properties, structure and bonding of these compounds have received major consideration. Spectroscopic properties of Cr^{4+} and Cr^{5+} ions doped in crystal lattices are also included for discussion. However, the formation of transient species in solution are generally not considered. Also, we have excluded the roles of chromium(IV) and chromium(V) species generated during chromium(VI) oxidation of organic[9, 10] and inorganic[11, 12] substrates. In this article chemistry of both the oxidation states are discussed together in each section dealing with principal types of compounds. Although no attempt has been made to make this review encyclopedic in nature, nevertheless care has been taken to include all major information that are available in literature up to middle of 1984. We note that some accounts of chromium(IV) and chromium(V) chemistry have appeared in the past[2, 13–15]. However, none of these are more recent that 10 years, and more importantly these articles have either dealt with certain special aspects or treated the subject as part of broader topics.

2 Oxides and Oxo-Compounds

2.1 Binary Oxide

CrO_2 is the only known binary oxide of chromium(IV). A great deal of information is available for CrO_2 because of its application in magnetic recording and storage devices, especially due to its superior properties compared to γ-Fe_2O_3. Pure CrO_2 can be prepared by thermal decomposition of CrO_3 under excess oxygen pressure[16, 17] or by hydrothermal decomposition of CrO_3[18–20]. Commercial methods[21–25] are based on hydrothermal reduction of chromium oxides having oxygen to chromium ratio greater than 2 · CrO_2 has been variously modified[22, 26–29] by incorporating different elements in order to improve magnetic properties, such as, coercivity, retentivity, saturation magnetization etc., as well as surface properties and morphology. Epitaxial growth[30] and thin film deposition[31] of this material has been reported.

CrO_2 has rutile structure (a = 4.421 Å, c = 2.917 Å: space group *P4/mnm*)[16, 19, 32, 33]. Four Cr–O distances are 1.918 Å and the other two Cr–O distances are slightly shorter, 1.882 Å; Cr–Cr separation is 2.918 Å. Anamolous temperature dependence of the lattice parameters of CrO_2 have been reported[20, 32, 33]. Of all known transition metal binary oxides CrO_2 is the only compound that exhibits room temperature ferromagnetic behavior. The ferromagnetic to paramagnetic transition temperature has been variously reported, lying in the range 389 to 398 K; however, the most recent values are 389 K[34] and 392 K[20]. Magnetic[19, 20, 27, 28, 34–36] and electrical properties[20, 35–40] of this material have been the subject of considerable research. The room temperature conductivity of CrO_2 single crystal is ca. $3 \times 10^3\ \Omega^{-1}\ cm^{-1}$ [20]. The metallic behavior of this compound, as well as its optical properties[41] can be explained in terms of a qualitative band model[42] suggesting that the partially filled conduction band results due to the overlap of the vacant d_{xy} orbital on the metal atom and filled P_π orbitals on the oxygen atom. There are two reports[43, 44] on XPS spectra of CrO_2. One such study[43] reports that the binding energies of core electron levels in CrO_2 are similar to that of Cr_2O_3 and CrOOH. Interestingly, the other report[44] claims that CrO_2 should be regarded as: $Cr_2O_3 \cdot CrO_3$.

No chromium(V) oxide has been reported so far. Compounds of apparent compositions Cr_3O_8 and Cr_2O_5 are known to be produced on controlled thermolysis of CrO_3 under positive oxygen pressure[17]. However, the phase Cr_2O_5 was always found to be oxygen deficient and approached the composition $CrO_{2.4}$. Subsequently single crystals of two oxides of composition Cr_5O_{12} and Cr_3O_8 were obtained[45]. X-ray crystal structure analysis of Cr_5O_{12} showed[46] that it should be formulated as $Cr_2^{III}(Cr^{VI}O_4)_3$. By analogy, Cr_3O_8 should be regarded as $Cr^{III}(Cr^{VI}O_4)_2$.

2.2 Ternary Oxides and Oxo-Compounds

A large number of compounds belongig to this category are known for both + 4 and + 5 oxidation states of chromium. Well-characterized tetravalent chromates of alkali and alkaline earth metals include: $M_4^ICrO_4$ (M^I = Na, Cs); $M_2^{II}CrO_4$ (M^{II} = Sr, Ba); Ba_3CrO_5; $M^{II}CrO_3$ (M^{II} = Ca, Sr, Ba, Pb).

Highly hygroscopic and reactive Na_4CrO_4[47] is obtained by heating a mixture of $NaCrO_2$ and Na_2O under vacuum

$$NaCrO_2 + 2\ Na_2O \xrightarrow{410°C} Na_4CrO_4 + Na\uparrow$$

The powder diffraction data of this compound has been indexed to an orthorhombic cell[47, 48] and it is probably isostructural with Na_4TiO_4 and Na_4VO_4 · Cs_4CrO_4 can be prepared by reducing Cs_2CrO_4 with two-equivalents of caesium[49]

$$Cs_2CrO_4(s) + 2\ Cs(l) \rightarrow Cs_4CrO_4$$

The standard enthalpy of formation of this compound, $-(1588.5 \pm 3.2)$ kJ mol^{-1}, determined calorimetrically[50], is in good agreement with the values $-(1550 \pm 140)$ and $-(1480 \pm 90)$ kJ mol^{-1} based on e.m.f. and vapor-presure measurements[49], respectively.

Ba_2CrO_4 and Sr_2CrO_4 are obtained[51,52] according to the reaction

$$Cr_2O_3 + M^{II}CrO_4 + 5\ M(OH)_2 \xrightarrow[N_2/Ar]{1000,\ 900\,°C} 3\ M_2^{I}CrO_4 + 5\ H_2O$$

There are other ways of preparing Ba_2CrO_4[7,53]:

$$BaCrO_4 + Ba(OH)_2 \xrightarrow{400\text{–}500\,°C} Ba_2CrO_4 + H_2O + \tfrac{1}{2}\ O_2$$

$$BaCrO_4 + Cr_2O_3 + 5\ Ba(OH)_2 \xrightarrow[N_2]{1000\,°C} 3\ Ba_2CrO_4 + 5\ H_2O$$

When the last reaction is carried out with excess of $Ba(OH)_2$, Ba_3CrO_5 is formed[51]

$$BaCrO_4 + Cr_2O_3 + 8\ Ba(OH)_2 \xrightarrow[N_2]{900\,°C} 3\ Ba_3CrO_5 + 8\ H_2O$$

The room temperature magnetic moment of Ba_2CrO_4 (2.82 B.M.)[53] is consistent with d^2 configuration of chromium(IV). Sr_2CrO_4 is orthorhombic with the space group $Pn2_1a$, and is structurally related to Ba_2TiO_4[52]. Sr_2CrO_4 has shown some interesting structural features[54]. There are eight formula units per unit cell and there are two types of CrO_4 tetrahedral sites. In each CrO_4 tetrahedra three Cr–O distances are similar (av. 1.84 and 1.90 Å), while the fourth bond (1.66 and 1.67 Å) is unusually short[54].

$CaCrO_3$[55], $SrCrO_3$[56] and $PbCrO_3$[57] have been obtained by reacting stoichiometric amounts of the metal oxides and CrO_2 at high temperature and pressure. $SrCrO_3$ and $PbCrO_3$ are cubic, but $CaCrO_3$ is orthorhombic. All of them have perovskite structure with the chromium atom octahedrally surrounded by six oxygen atoms. $CaCrO_3$ is a semiconductor[55] ($\sigma \approx 10^{-6}\ \Omega^{-1}\ cm^{-1}$ at 77 K; $10^{-4}\ \Omega^{-1}\ cm^{-1}$ at 300 K) and below the Neél temperature (90 K) exhibits parasaitic ferromagnetism. $SrCrO_3$ behaves as a metallic conductor ($\sigma \approx 10^5\ \Omega^{-1}\ cm^{-1}$ at 4 K; $2.5 \times 10^3\ \Omega^{-1}\ cm^{-1}$ at 298 K) and is a Pauli paramagnetic material[56]. Neutron diffraction data of $PbCrO_3$ at 77 and 4.2 K have indicated[57] the absence of any magnetic or structural phase transition.

Several polytypic modifications of $BaCrO_3$ have been obtained in the reaction between Ba_2CrO_4 and CrO_2 at 900 to 1300 °C and 60 to 65 kbar in a tetrahedral anvil press[58]. The optimum temperatures for preparing these polytypes are: 4 H, 1000 °C[59]; 6 H, 900 °C[60]; 14 H, 1300 °C[61]; 27 R, 1200 °C[62]. The most stable forms are 4 H and 6 H varieties. All these polytypes are structurally related, and may be described in terms of hexagonal (h) layers, for which the two neighbouring layers are alike, and the cubic (c) layers, for which the two neighbouring layers are of different types. According to this description, the polytypes have the following sequences:

4 H, *hchc*[59]; 6 H, *hcchcc*[60]; 14 H, *hchchcchchchcc*[61]; 27 R, *hchchcc*[62]

The "unit" of structure in these polytypes is the binuclear Cr_2O_9 face-sharing pair of CrO_6 octahedra. These octahedral pairs are corner linked to each other in 4 H (Fig. 1), but in 6 H the octahedral pairs are corner linked to two individual CrO_6 octahedra as shown in

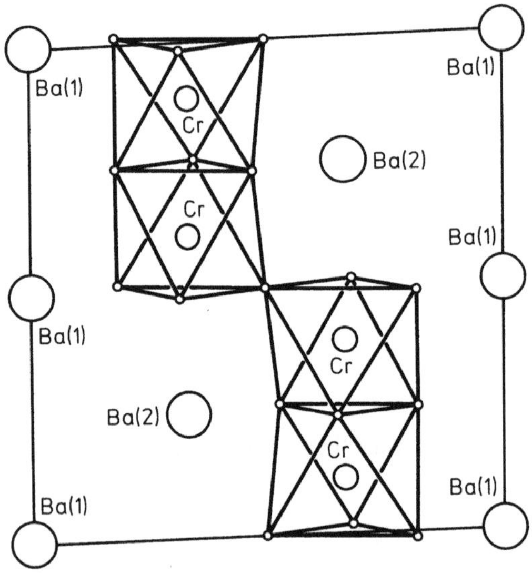

Fig. 1. Projection of the hexagonal (110) plane in 4 H $BaCrO_3$ polytype. Reproduced with permission from Ref. 59. Copyright 1982, Academic Press

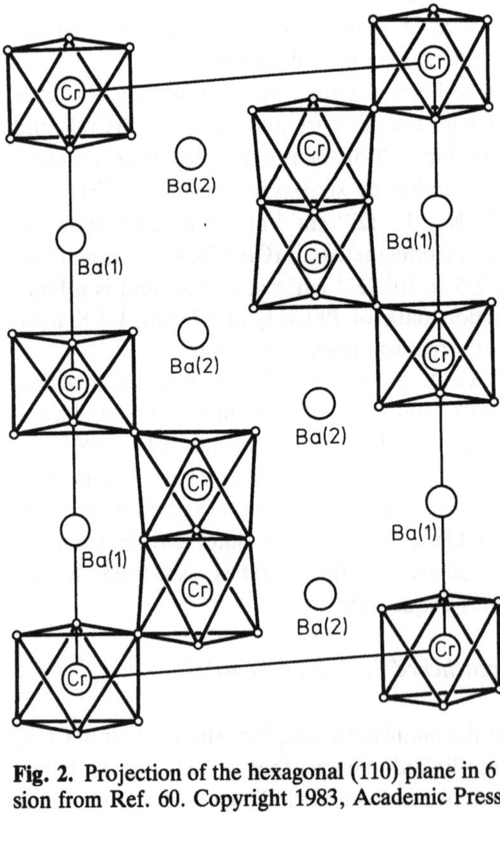

Fig. 2. Projection of the hexagonal (110) plane in 6 H $BaCrO_3$ polytype. Reproduced with permission from Ref. 60. Copyright 1983, Academic Press

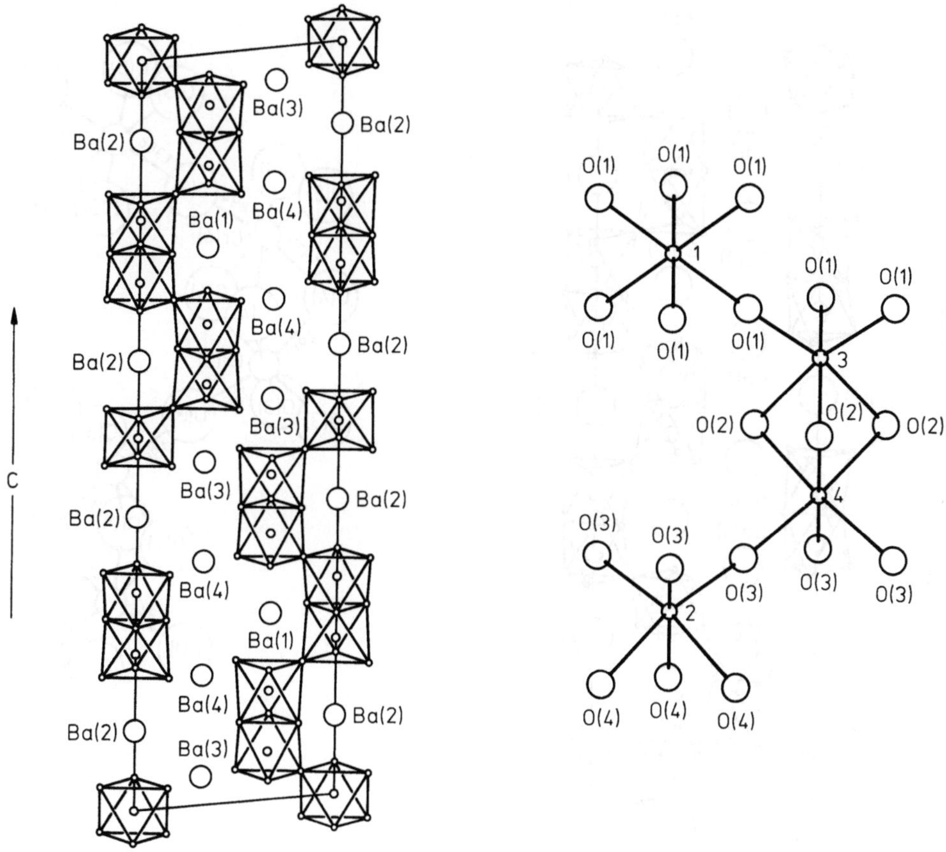

Fig. 3. a Projection of the hexagonal (110) plane in 14 H $BaCrO_3$ polytype. Reproduced with permission from Ref. 61. Copyright 1982, International Union of Crystallography. **b** The coordination of the four crystallographic types of chromium in 14 H $BaCrO_3$. Reproduced with permission from Ref. 61. Copyright 1982, International Union of Crystallography

Fig. 2. Figures 3 and 4 depict the structures of 14 H and 27 R polytypes, respectively, which have mixtures of 4 H and 6 H features.

Further structural consideration implied[62] that there is a "missing" polytype which should be designated as 15 R. In all these structures each of the barium atoms are surrounded by twelve oxygen atoms. The lattice images of 4 H $BaCrO_3$[63] are consistent with $(ch)_2$ description. Average Cr–Cr distances in $BaCrO_3$ polytypes are: 2.61 (4 H), 2.65 (6 H), 2.64 (14 H) and 2.63 Å (27 R); average ionic radius for Cr^{4+} ion is 0.55 Å. More detailed structural parameters of these polytypes are given in Table 2. $BaCrO_3$ polytypes show fairly high electrical conductivities and low activation energies[58]. Typically for the 4 H modification, $\sigma \approx 2.5\ \Omega^{-1}\ cm^{-1}$ at 300 K and $E^a \approx 0.11$ eV. Another phase previously considered as 12 R $BaCrO_3$[58] was later found to be[64] of non-stoichiometric composition, $Ba_2Cr_{7-x}O_{14}$ ($x \approx 0.5$); it contains chromium(III) octahedral and chromium(VI) tetrahedral sites.

Several alkali, alkaline earth and rare earth chromates(V) containing discrete CrO_4^{3-} tetrahedra are known. Examples are: $M_3^ICrO_4$ (M^I = Li, Na, K, Cs), $M_3^{II}(CrO_4)_2$ (M^{II} =

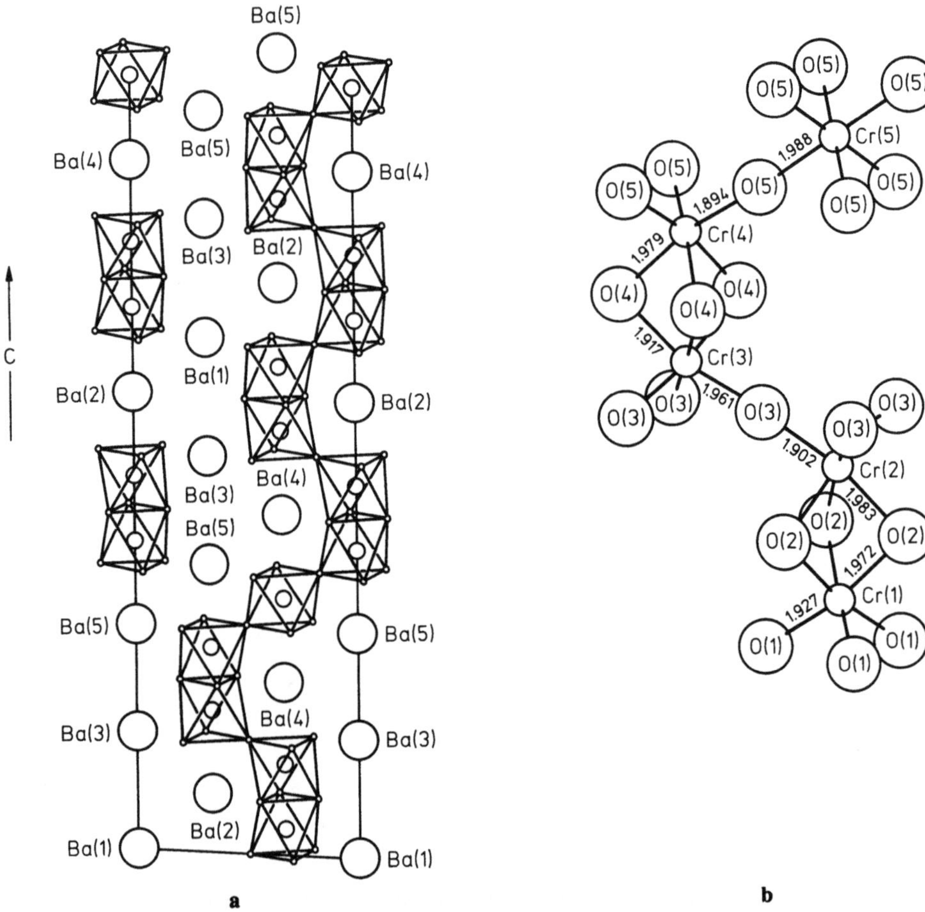

Fig. 4. **a** The structure of 27 R $BaCrO_3$ polytype. Reproduced with permission from Ref. 62. Copyright 1980, Academic Press. **b** The coordination of the five crystallographic types of chromium in 27 R $BaCrO_3$. Reproduced with permission from Ref. 62. Copyright 1980, Academic Press

Table 2. Crystal structure data of $BaCrO_3$ polytypes

Polytype	Structural parameters	Ref.
4 H	Hexagonal, a = 5.660, c = 9.357 Å; space group *P6₃/mmc;* Cr–O(1) × 3 = 1.934, Cr–O(2) × 3 = 1.952 Å; av. Ba–O = 2.87 Å	59
6 H	Hexagonal, a = 5.629, c = 13.698 Å; space group *P6₃/mmc;* Cr(1)–O(2) × 6 = 1.967, Cr(2)–O(1) × 3 = 1.941, Cr(2)–O(2) × 3 = 1.904 Å; av. Ba–O = 2.83 Å	60
14 H	Hexagonal, a = 5.650, c = 32.467 Å; space group *P6₃/mmc;* Cr(1)–O(1) × 6 = 1.985, Cr(2)–O(3) × 3 = 1.912, Cr(2)–O(4) × 3 = 1.993, Cr(3)–O(1) × 3 = 1.900, Cr(4)–O(2) × 3 = 1.928, Cr(4)–O(3) × 3 = 1.953 Å; av. Ba–O = 2.85 Å	61
27 R	Rhombohedral, a = 5.652, c = 62.75 Å (hexagonal cell dimensions); space group $R\bar{3}m$; Cr(1)–O(1) × 3 = 1.927, Cr(1)–O(2) × 3 = 1.972, Cr(2)–O(2) × 3 = 1.983, Cr(2)–O(3) × 3 = 1.906, Cr(3)–O(3) × 3 = 1.961, Cr(3)–O(4) × 3 = 1.916, Cr(4)–O(4) × 3 = 1.979, Cr(4)–O(5) = 1.894, Cr(5)–O(5) = 1.987 Å; av. Ba–O = 2.91 Å	62

Ca, Sr, Ba), $M^{II}_2(CrO_4)Cl$ (M^{II} = Ca, Sr), $M^{II}_5(CrO_4)_3X$ (M^{II} = Ca, Sr, Ba: X = OH, F, Cl) and $M^{III}CrO_4$ (M^{III} = rare earths and yttrium). Structurally these compounds are related to the corresponding phosphates, vanadates and silicates. Except compounds of the type $M^{I}_3CrO_4$, others are air-stable in the solid state. Pentavalent chromates disproportionate to chromium(VI) and chromium(III) in dilute acid solutions

$$3\,CrO_4^{3-} + 8\,H^+ \rightarrow 2\,CrO_4^{2-} + Cr^{3+} + 4\,H_2O$$

Oxygen and moisture sensitive $M^{I}_3CrO_4$ compounds have been prepared in several different ways. Li_3CrO_4 can be obtained by heating a mixture of Li_2CrO_4 and Li_2O or Li_2CO_3 at 600 to 800 °C under nitrogen atmosphere[65]. In another method[66] a mixture of Li_2CO_3 and Cr_2O_3 is heated in air at 800 °C for 12 h and then annealed under argon atmosphere

$$3\,Li_2CO_3 + Cr_2O_3 + O_2 \rightarrow 2\,Li_3CrO_4 + 3\,CO_2$$

Li_3CrO_4 is hexagonal and isomorphous with Li_3PO_4[66]. Na_3CrO_4 is formed according to the following reactions[48, 65, 67, 68]:

$$Na_2CrO_4 + NaN_3 \xrightarrow[N_2]{300°,\ 800\,°C} Na_3CrO_4 + 1.5\,N_2 \qquad 65$$

$$NaCrO_2 + Na_2O_2 \xrightarrow[N_2]{400°,\ 850\,°C} Na_3CrO_4 \qquad 65$$

$$4\,Na_2CrO_4 + Cr_2O_3 + 10\,NaOH \xrightarrow[N_2]{600\,°C} 6\,Na_3CrO_4 + 5\,H_2O \qquad 67$$

$$5\,Na_2O + Cr_2O_3 + 4\,Na_2CrO_4 \xrightarrow[Ar]{600°\text{–}800\,°C} 6\,Na_3CrO_4 \qquad 68$$

$$2\,Na_2CrO_4 + Na_2O \xrightarrow[Ar/Vacuum]{350°,\ 600\,°C} 2\,Na_3CrO_4 + O_2 \qquad 48$$

The preparative methods for K_3CrO_4 are[67, 69]:

$$Cr_2O_3 + 4\,K_2CrO_4 + 10\,KOH \xrightarrow[N_2]{450\,°C} 6\,K_3CrO_4 + 5\,H_2O \qquad 67$$

$$5\,K_2O + 4\,K_2CrO_4 + Cr_2O_3 \xrightarrow[Ar]{350°,\ 550\,°C} 6\,K_3CrO_4 \qquad 69$$

Cs_3CrO_4 has been prepared[70] by reducing Cs_2CrO_4 with one-equivalent of caesium

$$Cs_2CrO_4 + Cs \xrightarrow[Ar]{500°,\ 400\,°C} Cs_3CrO_4$$

Except Li_3CrO_4 other three alkali metal chromates(V) undergo phase transitions. The room temperature cubic form of Na_3CrO_4 transforms to a tetragonal phase at 300 °C[68]. The corresponding phase transition for K_3CrO_4 occurs at 180 °C[69]. Cs_3CrO_4 which is tetragonal at room temperature transforms to a phase of unknown symmetry at 500 °C[70]. The enthalpy and entropy changes for the phase transition of Cs_3CrO_4 are 1.8 kJ mol^{-1} and 2.3 JK^{-1} mol^{-1}, respectively[70]. Heat of formation of Cs_3CrO_4 is $-(1543.13 \pm 2.66)$ kJ mol^{-1}[70]. It may be noted that this value is very close to the heat of formation of Cs_4CrO_4 $-(1588.5 \pm 3.2)$ kJ mol^{-1}[50].

Among the alkaline earth chromates, $M_3^{II}(CrO_4)_2$, strontium and barium compounds are easily formed in a solid state reaction between the metal chromates and carbonates[66,71]. They are tetragonal and isostructural with $Ba_3(PO_4)_2$. The formation of $Ca_3(CrO_4)_2$ by heating $CaCrO_4$ with excess of $Ca(OH)_2$[72] or reacting stoichiometric amounts of $CaCO_3$ and Cr_2O_3 at 1000 °C[73,74] have been reported.

Chromium(V) analogs of the minerals apatite, $Ca_5(PO_4)_3F$, and spodiosite, $Ca_2(PO_4)F$ have been reported by various workers[66,71,74–77]. A few preparative methods are given here

$$3\ Ca_3(CrO_4)_2 + Ca(OH)_2 \xrightarrow[900\,°C]{\text{air, moisture}} 2\ Ca_5(CrO_4)_3OH$$

$$10\ MO + 3\ Cr_2O_3 \xrightarrow[900\,°C]{\text{air, moisture}} 2\ M_5(CrO_4)_3OH \quad (M = Ca, Sr, Ba)$$

$$9\ CaCO_3 + 3\ Cr_2O_3 + CaF_2 \xrightarrow[950\,°C/N_2]{900\,°C/\text{air;}} 2\ Ca_5(CrO_4)_3F$$

$$3\ Ca_3(CrO_4)_2 + CaCl_2 \xrightarrow[N_2]{900\text{–}950\,°C} 2\ Ca_5(CrO_4)_3Cl$$

$$9\,M(OH)_2 + 3\ Cr_2O_3 + MX_2 \xrightarrow[950\,°C/N_2]{900\,°C/\text{air;}} 2\ M_5(CrO_4)_3X \quad (M = Sr, Ba;\ X = Cl, F)$$

Crystals of $M_2^{II}(CrO_4)Cl$ (M = Ca, Sr) have been grown[66,74] by heating $M_3^{II}(CrO_4)_2$ in a flux of MCl_2 at 700 °C under inert atmosphere. All $M_5^{II}(CrO_4)_3X$ type of compounds crystallize with hexagonal unit cells and *P6₃/m* space group. The Cr–O distances in the CrO_4 tetrahedra of $Ca_5(CrO_4)_3OH$ lie in the range 1.64–1.68 Å[66]. X-ray studies of Ca_2CrO_4Cl and Ca_2PO_4Cl (both orthorhombic; space group *Pbcm*) have revealed[78] that these structures are made up to discrete CrO_4^{3-} and PO_4^{3-} tetrahedra which are held together by Ca^{2+} ions. However, in both the cases the CrO_4^{3-} and PO_4^{3-} tetrahedra are distorted from ideal geometry; the distortion being more with CrO_4^{3-} ($\sphericalangle$ OCrO = 105.1, 104.6, 119.1°) than with PO_4^{3-} ($\sphericalangle$ OPO = 107.3, 107.8, 113.6°). In Ca_2CrO_4Cl the Cr–O distances are 1.68 and 1.71 Å. Infrared spectra of Ca_2CrO_4Cl, Ca_2VO_4Cl and Ca_2AsO_4Cl also indicated[79] greater tetrahedral distortion for CrO_4^{3-}.

Rare earth chromates(V) are air-stable compounds. Two convenient preparative methods are[80,81]

$$Ln_2(CrO_4)_3 + Ln(NO_3)_3 \xrightarrow{600\,°C/12\,h} 3\,LnCrO_4 + 3\,NO_2 + 1.5\ O_2 \qquad 80$$

$$LnCr(C_2O_4)_3 \cdot n\ H_2O + 2\ O_2 \xrightarrow[1\ h/air]{500-600\,°C} LnCrO_4 + 6\,CO_2 + n\ H_2O \qquad 81$$

$LaCrO_4$ is monoclinic and have monazite ($ThSiO_4$) structure[80]. $NdCrO_4$ and other higher members are tetragonal with Zircon ($ZrSiO_4$) structure[80]. $PrCrO_4$ is dimorphic[80–82], and $CeCrO_4$ could not be prepared. $La_{1-x}Nd_xCrO_4$ solid solutions exist in the monoclinic phase for x up to 0.23[81].

Crystal structure analysis for $YCrO_4$ (tetragonal, space group *I4₁/amd*) has been made[83]; the shortest Cr–O distance is 1.66 Å and the shortest Y–O distance is 2.44 Å. Activation energies for the thermal decomposition reaction have been determined[84]. The enthalpy change for

$$LnCrO_4 \rightarrow LnCrO_3 + 0.5\,O_2$$

the decomposition reaction is ca. 16 kJ mol^{-1}[84].

Magnetic measurements carried out for various pentavalent chromates invariably show moments close to spin-only value for a single unpaired electron (1.73 B.M.). Considerable attention has been focussed on ground state and excited state configurations of Cr^{5+} in CrO_4^{3-}. For determining the ground state EPR spectra of CrO_4^{3-} doped in single crystals of various isostructural compounds have been investigated[79, 85–91]. In pure tetrahedral symmetry, the d_{z^2} and $d_{x^2-y^2}$ orbitals are degenerate and lowest in energy. With a small distortion into D_{2d} symmetry one of these orbitals become lowest in energy. The fact that the EPR spectra of CrO_4^{3-} are too broad to be observed at room temperature[92, 93] is an indication that the distortion is not large, because the close proximity of an empty orbital to the ground state would lead to short relaxation times and hence broadening of lines at higher temperatures[79]. It can be shown from crystal field theory that "squashing" of the tetrahedron along the Z-axis would make the d_{z^2} orbital to lie lowest. On the other hand $d_{x^2-y^2}$ orbital will lie lowest if elongation of an S_4 axis occurs for a tetrahedron. By far the majority of cases EPR studies have conclusively shown that the unpaired electron of CrO_4^{3-} in various host lattices occupy the d_{z^2} ground state. Although on the structural ground CrO_4^{3-} doped in YPO_4 and YVO_4 lattices were expected to show a $d_{x^2-y^2}$ ground state, but a d_{z^2} is observed[90, 91]. This inversion is attributed to strong covalency effect[90]. The EPR spectra of CrO_4^{3-} doped in $NaVO_3$ crystals, however, have been interpreted on the basis of $d_{x^2-y^2}$ ground state[94]. Definite occupancy of $d_{x^2-y^2}$ orbital for CrO_4^{3-} doped in ferroelectric materials such as KH_2AsO_4, KH_2PO_4 etc. have been reported[95–99]. More details are given in Sect. 10.

Low temperature optical absorption spectra of CrO_4^{3-} in different host lattices have also been investigated in considerable detail[15, 79, 101–104]. $Ca_2(CrO_4, PO_4)Cl$[100, 101], $Ca_2(CrO_4, VO_4)Cl$[102] and $Sr_2(CrO_4, VO_4)Cl$[102] have similar spectral features and have been explained in terms of the energy level diagram[101] shown in Fig. 5. The manifolds at 10000, 17000 and 27000 cm^{-1} have been assigned to $^2E(t_1^6 2e^1) \rightarrow {}^2T_2(t_1^6 2e^0 t_2^1)$, $^2T_1(t_1^5 2e^2)$ and $^2T_2(t_1^5 2e^2)$ transitions, respectively. Crystal field separation energies in these systems are ca. 10000 cm^{-1}. However, another report[103] considers the absorption at 17000 cm^{-1} as low-symmetry component of the $^2T_2$ ligand field transition. Conflicting band assignments have been reported for $Ca_5[(CrO_4)_3, (PO_4)_3]X$ (X = F, Cl)[15] and $Sr_5[(CrO_4)_3, (PO_4)_3]Cl$[104]. While in the calcium apatites the absorptions occuring at ca. 9100 cm^{-1} are considered[15] due to $^2E \rightarrow {}^2T_2$ transition and those appearing at ca. 13000, 17000 and

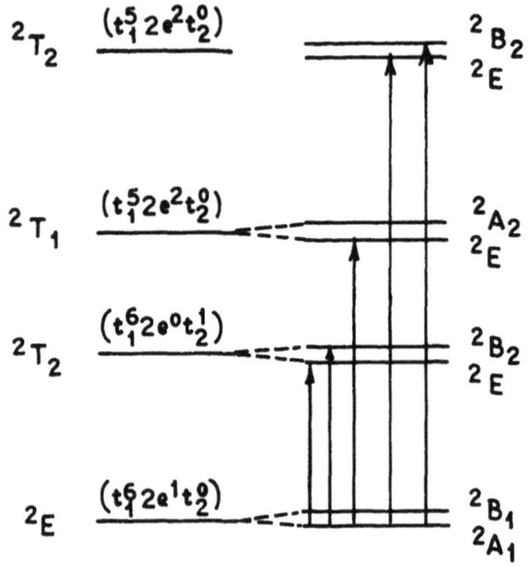

Fig. 5. The energy level diagram for CrO_4^{3-} [101]

21000 cm^{-1} due to charge transfer transitions, however, in the case of strontium analogue the absorptions occuring in the region 10000 to 17000 cm^{-1} are all considered[104)] due to ligand field transitions.

3 Peroxo Compounds

Quite a few well-characterized chromium(IV) peroxo compounds are known. These are: $[Cr(NH_3)_3(O_2)_2]$, $[Cr(en)(H_2O)(O_2)_2] \cdot H_2O$, $[Cr(pn)(H_2O)(O_2)_2] \cdot 2\,H_2O$, $[Cr(en)(NH_3)(O_2)_2] \cdot H_2O$, $[Cr(dmen)(H_2O)(O_2)_2] \cdot H_2O$, $[Cr(dien)(O_2)_2] \cdot H_2O$ and $K_3[Cr(CN)_3(O_2)_2]$ (en = 1,2-diaminoethane: pn = 1,2-diaminopropane; dmen = 2-(dimethyl)1,2-diaminoethane; dien = diethylenetriamine). A hexamethylenediamine derivative reported earlier[105)], however, could not be obtained in pure form later[106)]. Attempts to prepare peroxo compounds using various other primary, secondary and tertiary amines, as well as diamines, as the secondary ligands were also unsuccessful[106)].

$[Cr(NH_3)_3(O_2)_2]$ was initially prepared in several different ways[3, 107)], but the most convenient procedure[108)] is based on heating a reaction mixture containing $(NH_4)_3[Cr(O_2)_4]$ to 50 °C, followed by cooling at 0 °C. $[Cr(en)(H_2O)(O_2)_2] \cdot H_2O$[109)] and the diamino analogs[106, 110, 111)] are obtained by treating an aqueous solution of CrO_3 and diamine with H_2O_2 at 0 °C. $[Cr(en)(NH_3)(O_2)_2] \cdot H_2O$ is prepared by dissolving $[Cr(en)(H_2O)(O_2)_2] \cdot H_2O$ in 8 M NH_3[106)], and $K_3[Cr(CN)_3(O_2)_2]$ results on dissolution of $[Cr(NH_3)_3(O_2)_2]$ in aqueous KCN solution[112)].

In general these compounds are stable in solid state at the ambient temperature or below, but explode violently on heating. The room temperature magnetic moments of these compounds (ca. 2.8 B.M.)[113, 114)] are consistent with the presence of two unpaired electrons. They are also characterized by one or more sharp bands between 860 and 890 cm^{-1}[113, 115)] in their IR spectra. $[Cr(NH_3)_3(O_2)_2]$ was conceived much earlier[116)] to be a useful precursor for generating amino complexes of chromium(III). In recent years various chromium(III) complexes which otherwise are inaccessable have been prepared from chromium(IV) diperoxo compounds[106, 109, 111, 117, 118]. A few examples are given here

$$[Cr(en)(H_2O)(O_2)_2] \xrightarrow{HCl} [Cr(en)(H_2O)_2Cl_2]Cl$$

$$[Cr(en)(H_2O)(O_2)_2] \xrightarrow{HF + NH_4HF_2} NH_4[Cr(en)F_4]$$

$$[Cr(dien)(O_2)_2] \xrightarrow{HCl} [Cr(dien)Cl_3]$$

$$[Cr(dien)(O_2)_2] \xrightarrow{HF} [Cr(dien)(H_2O)_3]^{3+}$$

X-ray crystal structures of $[Cr(NH_3)_3(O_2)_2]$[119], $[Cr(en)(H_2O)(O_2)_2] \cdot H_2O$[120] and $K_3[Cr(CN)_3(O_2)_2]$[121] have been determined. In all cases chromium form a deformed pentagonal bipyramid. $[Cr(NH_3)_3(O_2)_2]$ crystallizes in rhombic as well as monoclinic forms. A two-dimensional X-ray analysis of the monoclinic form[122] gave basically same molecular structure as obtained with the rhombic form, but structural parameters are less precise. Structural data collected for the chromium(IV) peroxo compounds in Table 3 show that the Cr–O distances in these compounds are almost equal (1.87–1.91 Å), and the O–O distances (ca. 1.45 Å) are similar to those observed in other peroxides.

Unlike chromium(IV) compounds, only tetraperoxychromates(V), $M^I_3[Cr(O_2)_4]$ (M = alkali metals) are known. $M^I_3[Cr(O_2)_4]$ compounds are still prepared according to the original method[4] by treating a cold alkaline solution of $M^I_2CrO_4$ with 30% H_2O_2 which is then cooled below 0 °C. In solid state they are indefinitely stable at room temperature. In neutral or alkaline solution $[Cr(O_2)_4]^{3-}$ gets oxidized to CrO_4^{2-}, while in

Table 3. Crystal structure data of chromium(IV) peroxo complexes

Compound	Structural detail and comments	Ref.
$[Cr(NH_3)_3(O_2)_2]$	The orthorhombic form has the space group *Pnma*. The seven atoms around Cr^{4+} form a deformed pentagonal bipyramid. Chromium atoms can be placed in 8- and 4-fold positions. The molecules are held together by hydrogen bonds. 8-fold position: Cr–O = 1.86, 1.87, 1.88, 1.89 Å; Cr–N = 2.08, 2.11, 2.15 Å; O–O = 1.41, 1.43 Å. 4-fold position: Cr–O = 1.87, 1.91 Å; Cr–N = 2.08, 2.14, 2.16 Å; O–O = 1.45 Å.	119
$[Cr(en)(H_2O)(O_2)_2] \cdot H_2O$	Orthorhombic, space group *Pbc2₁*. One N of en and two O_2 groups are coplanar with Cr, the other N of en and H_2O molecule occupy axial positions. Cr–O(peroxo) = 1.86, 1.91 Å; Cr–N = 2.05, 2.07 Å; Cr–O(aqua) = 1.91 Å; O–O = 1.46 Å.	120
$K_3[Cr(CN)_3(O_2)_2]$	Monoclinic, space group *P2/a*. Two peroxo groups and one cyano group form the base of the pyramid and two other cyano groups occupy the apical positions. Cr–O = 1.88, 1.90 Å; Cr–C(cyano) = 2.08, 2.09, 2.11 Å; O–O = 1.45 Å.	121

acidic solution reduction to Cr^{3+} occurs. There is evidence[123] that in basic aqueous solution an equilibrium exists between tetraperoxychromate(V) and diperoxychromate(VI). The equilibrium constant for the reaction measured spectrophotometrically is pK = 12.6(0.2 M $NaClO_4$)[123].

$$[Cr^{VI}(O_2)_2(O)(OH)]^- + 1.5\,H_2O_2 \overset{K}{\rightleftharpoons} [Cr^{V}(O_2)_4]^{3-} + 2\,H^+ + H_2O$$

It is believed that singlet molecular oxygen, $O_2(^1\Delta g)$ is liberated from aqueous solution of $K_3[Cr(O_2)_4]$

$$4\,Cr(O_2)_4^{3-} + 2\,H_2O \rightarrow 4\,CrO_4^{2-} + 7\,O_2 + 4\,OH^-$$

The presence of singlet oxygen has been verified[124] by monitoring the emission at 1270 nm due to

$$O_2(^1\Delta g) \rightarrow O_2(^3\Sigma g^-) + \hbar\nu$$

Oxidation of olefines by K_3CrO_8 is thought to be carried out by singlet oxygen[124].

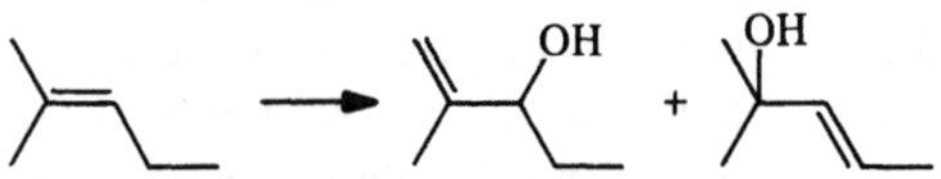

$K_3[Cr(O_2)_4]$ is isomorphous with $K_3[Nb(O_2)_4]$ and $K_3[Ta(O_2)_4]$. X-ray crystal structure determined for $K_3[Cr(O_2)_4]$ shows[125] that the chromium atom is surrounded by four peroxide ions in a distorted dodecahedral environment. Refinement (R = 8.5%)[126] of these X-ray data[125] reveals that the two Cr–O distances are significantly different ($Cr–O_I$ = 1.874, $Cr–O_{II}$ = 1.972 Å) and O–O distance (1.472 Å) is close to that found in alkali peroxides (1.49 Å). A different refinement (R = 10%)[127] with the same set of X-ray data[125], however, gives $Cr–O_I$ = 1.846, $Cr–O_{II}$ = 1.944 Å and O–O = 1.405 Å. Shortening of one Cr–O distance relative to the other (note the difference with chromium(IV) compounds) has been rationalized[127] by taking into consideration strong covalent interaction in the molecule. A SCF-MO-LCAO calculation carried out for CrO_8^{3-} ion shows[128] that bulk of the Cr–O bonding arises from overlap of chromium atomic orbitals with O–O in-plane π-bonding density and in-plane π-antibonding density.

The infrared spectra of $M_3^I[Cr(O_2)_4]$ compounds[113, 129] show normal peroxo linkage, and their room temperature magnetic moment[130, 131] give spin-only values (ca. 1.8 B.M.). The EPR spectrum of polycrystalline K_3CrO_8 ($g_{\parallel}$ = 1.936, $g_{\perp}$ = 1.983)[127] is in good agreement with the spectrum of K_3CrO_8 doped in K_3NbO_8 ($g_{\parallel}$ = 1.944, $g_{\perp}$ = 1.985)[131a]. The electronic spectrum of K_3CrO_8 in 1 M KOH solution gives two weak absorptions ($\varepsilon \simeq 50\ M^{-1}\,cm^{-1}$) at 16900 and 18000 cm^{-1} and are assigned[127] due to $d_{x^2-y^2} \rightarrow d_{z^2}$ and $d_{x^2-y^2} \rightarrow d_{xy,yz}$ transitions, respectively. On this basis the ligand field separation energy turns out to be 35000 cm^{-1}. The electronic spectrum of K_3CrO_8 dissolved in 30% H_2O_2, however, gives a single absorption band at 20000 cm^{-1} with strong intensity ($\varepsilon \simeq 500\ M^{-1}\,cm^{-1}$)[131a].

4 Halides and Halo Complexes

Among the binary halides of chromium(IV) CrF_4 exists in the solid state, others are stable only in gaseous state in the temperature range 800–1300 K. Several salts of the fluoro complex anions $[CrF_5]^-$, $[CrF_6]^{2-}$ and $[CrF_7]^{3-}$ have been reported.

CrF_4 is formed along with a small quantity of CrF_5 by direct fluorination of Cr, $CrCl_3$ or CrF_3 at 300–350 °C[132–134]. It is purified by vacuum sublimation at 150 °C. CrF_4 is a dark greenish-black solid, highly moisture sensitive, and apparently exists in amorphous form. The magnetic moment of CrF_4 (3.02 at 294 K and $\theta = -70°$)[134] probably indicates significant spin-orbit interaction. The reaction equilibrium

$$CrX_3(s) + 0.5\,X_2(g) \rightleftharpoons CrX_4(g)$$

for X = Cl, Br and I has been investigated by effusion, transpiration and spectrophotometric methods[135–139]. Table 4 summarizes the values of standard enthalpies of formation ($\Delta H°$), standard entropies ($\Delta S°$) and bond energies of CrX_4 (X = Cl, Br, I)[135, 138, 139]. Raman Spectral studies[140] have established tetrahedral structure for CrX_4 compounds. The stretching force constants of Cr–X bonds are considerably greater with $CrCl_4$ (266 Nm^{-1}) than $CrBr_4$ (208 Nm^{-1})[140]. The electronic spectrum of $CrCl_4$[138] has been interpreted on the basis of multiple scattering X_αMO calculations[141]. In T_d configuration the ground state for CrX_4 is $^3A_2$. The d–d bands observed at 12555, 10110 and 7050 cm^{-1} are considered due to the transitions from $^3A_2$ to $^3T_1(1)$, $^3T_2$ and $^1E(1)$ states, respectively. Another absorption band at 16485 cm^{-1} probably arises due to combination of $^3A_2 \rightarrow {}^1A_1$, $^1T_2(1)$ and $^1T_1$ transitions. The absorptions observed at 25000 and 30770 cm^{-1} are due to charge transfer transitions.

Compounds of composition $M^ICrF_5 \cdot 0.5\,BrF_3$ (M = K, Rb, Cs) have been obtained[134] by heating 1 : 1 mixture of CrF_4 and MCl in BrF_3. Removal of BrF_3 at 100 °C affords $M^ICrF_5 \cdot 0.5\,BrF_3$, but at 160 °C produces M^ICrF_5. In analogous way $M^I_2CrF_6 \cdot 0.5\,BrF_3$ and $M^I_2CrF_6$ have been obtained[134] by using 1 : 2 molar ratio of the reactants. More conveniently $M^I_2[CrF_6]$compounds are obtained by solid state fluorination of MCl and $CrCl_3$ (2 : 1) at 300 °C[133, 142]. $K_2[CrF_6]$ and $Rb_2[CrF_6]$ are dimorphic. The hexagonal form transforms to the cubic phase at a higher temperature[142]. $Cs_2[CrF_6]$ is cubic. Compounds of the type $M^{II}[CrF_6]$ (M^{II} = Sr, Ba, Ca, Cd, Hg) have been prepared by solid state fluorination reaction at 300 °C and 400 atm pressure for 2 to 3 days[143]. $SrCrF_6$ and $BaCrF_6$ are yellow colored compounds and isotypic to $Ba[GeF_6]$. Other $M^{II}[CrF_6]$ (M = Ca, Cd, Hg) compounds are rose colored and have $LiSbF_6$ structure. Recently $[NO]_2[CrF_6]$ has been prepared[144] by reacting NOF with CrF_5.

Table 4. Thermodynamic parameters of chromium tetrahalides

Compound	$\Delta H°$ (kcal mol^{-1})	$\Delta S°$ (cal mol^{-1} deg^{-1})	E (kcal mol^{-1})
$CrCl_4$	−101.6	87.5	77.8
$CrBr_4$	− 42.6	106.9	60.9
CrI_4	4.1	112.3	48

In $[CrF_6]^{2-}$ the chromium(IV) has an octahedral environment and the electronic ground state is $^3T_{1g}$. The electronic spectra of $M^I_2[CrF_6]$ (M = K, Rb, Cs) have similar features at room temperature and 77 K[145]. A prominent absorption at 20200 cm^{-1} is assigned to the spin allowed d–d transition $^3T_{1g}(t^2_{2g}) \rightarrow {}^3T_{2g}(t_{2g}e_g)$, and a very weak absorption at 11400 cm^{-1} is considered due to $^3T_{1g} \rightarrow {}^1E_g$, $^1T_{2g}$ transitions. The $^3T_{1g}(t^2_{2g}) \rightarrow {}^3T_{1g}(t_{2g}e_g)$ transition occurs at about 28400 cm^{-1}. In addition, two charge transfer transitions occur at 37000 and 30000 cm^{-1}. The ligand field parameters evaluated for $[CrF_6]^{-2}$ are: 10 Dq = 21700 cm^{-1}, B = 680 cm^{-1} and $\beta = 0.67$[145].

Rb_3CrF_7 and Cs_3CrF_7 have been prepared by solid state reaction between $M^I_2CrF_6$ and MF at 300 °C for several days[146]. They are isostructural with $M^I_3SiF_7$ (M = NH_4, K)[147, 148]. The structure of $(NH_4)_3SiF_7$ consists of an ordered aggregate of ammonium, fluoride, and regular octahedral SiF_6^{2-} ions[147].

CrF_5 is the only known halo compound of chromium(V). It is prepared by fluorination of chromium powder at 400 °C under 200 atm pressure[149]. CrF_5 is a crimson coloured volatile compound, melting at 30 °C. It is highly reactive and readily gets hydrolyzed by water. CrF_5 oxidizes PF_3, AsF_3 and SbF_3 to their corresponding pentafluorides[150], and forms adducts with CsF, SbF_5 and NO_2F[151]. CrF_5 is orthorhombic[152] and isostructural with VF_5, TeF_5 and ReF_5. It may be noted that the crystal structure of VF_5 consists of polymeric chains with *cis*-bridging of octahedrally disposed fluorine atoms[153]. Infrared and Raman spectra of CrF_5 in liquid state is consistent with *cis*-fluorine bridged polymeric structure[154].

5 Oxyhalides and Oxyhalo Complexes

Compounds of this category do not exist for tetravalent chromium. As regards chromium(V), a good deal of information is available with $CrOF_3$, $CrOCl_3$ and various salts of the complex anions $[CrOCl_4]^-$, $[CrOCl_5]^{2-}$, $[CrOF_4]^-$ and $[CrOF_5]^{2-}$.

Compounds of composition $CrOF_3 \cdot 0.25\ BrF_3$, $CrOF_3 \cdot 0.25\ BrF_5$ and $CrOF_3 \cdot 0.3\ ClF_3$ have been obtained[155] by reacting CrO_3 with BrF_3, BrF_5 and ClF_3, respectively. A similar reaction between ClF and CrO_3 or CrO_2F_2 affords a compound of composition $CrOF_3 \cdot$ nClF which on multiple treatment with fluorine at 120 °C produces unsolvated $CrOF_3$[156]. $CrOF_3$ is a bright purple solid which readily gets hydrolyzed to Cr^{3+} and CrO_4^{2-} ions[155]. At 500 °C $CrOF_3$ decomposes to CrF_3 and oxygen. It reacts with fluorine at 190 °C to form CrF_5, and a reaction with KF in presence of HF results in the formation of $KCrOF_4$[156]. The room temperature magnetic moment of $CrOF_3$ is 1.82 B.M.[155]. A polymeric structure for $CrOF_3$ with an octahedral geometry of chromium(V) has been proposed[156] on the basis of the IR and Raman spectral bands due to νCr=O (1000 cm^{-1}), νCr–F (740–600 cm^{-1}) and νCr–F–Cr (565 cm^{-1}).

$CrOCl_3$ has been prepared by reacting CrO_3 with SO_2Cl_2 or $SOCl_2$[157]; also by vacuum sublimation of the product obtained by treating CrO_2Cl_2 with BCl_3 at – 20 °C[158, 159]. It is a red-black solid, unstable above 0 °C. The magnetic moment of $CrOCl_3$ is close to spin-only value[158] and it gives an isotropic EPR spectrum in CCl_3F solution ($\langle g \rangle = 1.989$)[160]. Infrared and electronic spectra of matrix-isolated $CrOCl_3$ have been investigated[160]. In argon-matrix the stretching frequencies for the most abundant species, $^{52}Cr O^{35}Cl_3$ are observed at 1018 (νCr=O), 462 (E, νCr–Cl) and 410 cm^{-1} (A_1, νCr–Cl). Solid $CrOCl_3$, however, gives four bands at 1024, 435, 408 and 333 cm^{-1} [160]. Significant shift of νCr–Cl at lower frequencies suggests a polymeric structure for solid $CrOCl_3$ (similar to $CrOF_3$)[160]. It is likely that the structure of $CrOCl_3$ is close to that of $MoOCl_3$ whose crystal structure shows[161] an infinite chain with *cis*-dichloro bridging. Four peaks at 10.0(a), 11.4–12.0(b), 12.4–13.4(c) and 13.8–15.0(d) eV have been observed in He(I) and He(II) photoelectron spectra of $CrOCl_3$[162]. These are identified due to the ionization of Cr_{3d}(a), Cl_{3p}(b, c) and O_{2p}(d) electrons. The ionization energies calculated by configuration interaction (CI) and scattered wave (SW)X_α methods are in good agreement with the observed spectra[162]. An *ab initio* calculation[163] gives ^2E as the ground state of the molecule. The electronic spectrum of $CrOCl_3$ in Ar matrix[160] exhibits prominent bands at 40816, 31250, 27700(sh), 22133 and 20000(sh), and a weak band at 12970 cm^{-1}. The bands at 12970 and 20000 cm^{-1} are considered due to d–d transitions, and the remaining others due to Cl → Cr and O → Cr charge transfer transitions. The electronic spectrum of $CrOCl_3$ in acetic acid has been reported[164] to contain two absorption bands at 25640 and 19160 cm^{-1}.

With a variety of counter cations basically two types of oxyhalide complexes $[CrOX_4]^-$ and $[CrOX_5]^{2-}$ (X = F, Cl) have been isolated. Although a $[CrOBr_5]^{2-}$ species was reported[165], however, a subsequent study[166] failed to authenticate it. The oxyfluoro complexes are less numerous; known examples are: $K[CrOF_4]$, $Ag[CrOF_4]$ and $[Et_4N]_2[CrOF_5]$. Largest number compounds of the type $A[CrOCl_4]$ have been isolated with A^+ = Hpy, Hquin, Me_4N, Et_4N, Pr_4N, $AsPh_4$, $BzPPh_3$, PCl_4. The cations that stabilize $[CrOCl_5]^{2-}$ are K, Cs, H_2(2,2′)-bipy and H_2(4,4′)-bipy.

$M[CrOF_4]$ can be prepared by the action of BrF_3 on M_2CrO_4[167]. A similar reaction of BrF_5 with $K_2Cr_2O_7$ results $K[CrOF_4] \cdot 0.5\ BrF_5$[155]. The reaction between KF and $CrOF_3$ also yields $K[CrOF_4]$[155]. $K[CrOF_4]$ is orthorhombic[155] and has a room temperature magnetic moment of 1.8 B.M.[168].

$[Et_4N]_2[CrOF_5]$ has been prepared[169] by treating a HF solution containing AgF with a solution of $[Et_4N]_2[CrOCl_5]$ in CH_2Cl_2.

Salts of the oxychloro complex anions $[CrOCl_4]^-$ and $[CrOCl_5]^{2-}$ are prepared[166,169] by some modification of the original method due to Weinland and coworkers[5]. The method consists of adding an appropriate base or chloride salt of the desired cation to a glacial acetic acid solution of CrO_3 saturated with HCl. Further saturation of the mixture with HCl yields the desired complex. Highly moisture sensitive $[PCl_4][CrOCl_4]$ has been obtained[170] by reacting CrO_2Cl_2 with PCl_5. Dehydrohalogenation of $[H_2bipy][CrOCl_5]$ and $[H_2phen][CrOCl_5]$ in inert atmosphere has been reported[171] to produce $CrO(bipy)Cl_3$ and $CrO(phen)Cl_3$.

Of all the oxyhalo complexes of chromium(V), complete X-ray structural analysis has been made only for $[Ph_4As][CrOCl_4]$[172]. The square pyramidal $[CrOCl_4]^-$ ion has crystallographic *4 mm* (C_{4v}) symmetry with Cr–O = 1.519 Å, Cr–Cl = 2.240 Å and ∢O–Cr–Cl = 104.5°. The Cr–O distance is very short: a pπ-dπ bonding due to the overlap of the oxygen pπ orbital with unoccupied metal $d_{xz,yz}$ orbitals probably corresponds to a bond order of three[172]. $Cs_2[CrOCl_5]$ is cubic, isomorphous with $Cs_2[MOCl_5]$ (M = Nb, Mo, W)[174] and has an octahedral $K_2[PtCl_6]$ structure[173]. Magnetic measurement of this compound gives μ_{eff} = 1.8 B.M. and θ = − 14 K[173].

The oxychloro complexes of chromium(V) can be easily characterized from their characteristic Cr–O and Cr–Cl stretching frequencies (Table 5). It may be noted that while in $[CrOCl_4]^-$ ν(Cr–O) appears at ca. 1020 cm^{-1}, this band is shifted to about 930 cm^{-1} in $[CrOCl_5]^{2-}$. In several $[A^+][CrOCl_4^-]$ type of complexes an additional ν(Cr=O) vibration has been reported[169] at ca. 950 cm^{-1}. The origin of this band is not certain, it may as well be due to some CrO_2Cl_2 impurity. Infrared data of $M^I[CrOF_4]$ compounds are not available, however, in $[Et_4N]_2[CrOF_5]$, ν(Cr=O) appears at 960 cm^{-1} and ν(Cr–F) at 480 and 450 cm^{-1} [169]. Room temperature magnetic moments of the oxychloro complexes of chromium(V) (Table 5) are found to lie in the range 1.73 to 1.92 B.M.

There remains considerable interest in the electronic structures of chromium(V) oxyhalo compounds. Since the initial semiimperical MO calculations[175, 176] further modifications have been made[177–179], and more recent work include *ab initio* SCF–MO[180], *ab initio* CI[181] and SCF–MS–X_α methods[182, 183]. The ground state has turned out to be $^2B_2$ for both $[CrOX_4]^-$ and $[CrOX_5]^{2-}$ species from different calculations, the unpaired electron is an orbital of b_2 symmetry which is primarily a metal d_{xy} orbital. The relative energies of the empty orbitals, however, differ in different methods of calculations. For example, on the basis of semiemperical MO[175] and SCF–MS–X_α calculations[182, 183] the decreased order of the orbital energies is: $d_{xy}(b_2) < d_{xz,yz}(e) < d_{x^2-y^2}(b_1) < d_{z^2}(a_1)$. However, this order is at variance with the sequence $d_{xy} < d_{x^2-y^2} < d_{z^2} < d_{xy,yz}$ obtained from CI calculations[181] or $d_{xy} < d_{xz,yz} < d_{z^2} < d_{x^2-y^2}$ obtained from extended Hückel calculations[183].

Table 5. IR and magnetic data for the salts of $[CrOCl_4]^-$ and $[CrOCl_5]^{2-}$ complex anions

Compound	ν(Cr=O) cm^{-1}	ν(Cr–Cl) cm^{-1}	μ_{eff} B.M.	Ref.
$[Hpy][CrOCl_4]$	1019	398, 382, 343(sh)		166
$[Hquin][CrOCl_4]$	1016	397, 347(sh)		166
$[Hi\text{-}quin][CrOCl_4]$	1016	398, 348(sh)		166
$[Me_4N][CrOCl_4]$	1016	408(sh), ca. 380(br)		166
		400	1.79	169
$[Et_4N][CrOCl_4]$	1022	395, 345		166
	1035	395	1.78	169
$[Pr_4N][CrOCl_4]$	1005	400	1.74	169
$[Bu_4P][CrOCl_4]$	1025	385	1.74	169
$[BzPh_3P][CrOCl_4]$	1010	415	1.73	169
$[Cl_4P][CrOCl_4]$	1017	401, 348	1.91	170
$[Ph_4As][CrOCl_4]$	1019	398		166
	1020	405		169
$[H_2(2,2\text{-})bipy][CrOCl_5]$		366		166
$(2,2\text{-})bipyCrOCl_3$	950		1.85	171
$[H_2(4,4\text{-})bipy][CrOCl_5]$		354		166
$Rb_2[CrOCl_5]$	945	320		169
$Cs_2[CrOCl_5]$	927	334, 312(sh)		166
	945	320	1.85	169
			1.80	174

py = pyridine; quin = quinoline; i = quin = iso-quinoline; Bz = benzyl; bipy = bipyridine

The electronic spectral features of known $[CrOX_4]^-$ and $[CrOX_5]^{2-}$ species are shown in Table 6. For $[Ph_4As][CrOCl_4]$ alone polarized spectra has been recorded[181] at 4 K. In other cases mull spectra have been recorded either at 77 K or at the ambient temperature[166, 169]. For many compounds solution spectra are also available[166, 169]. In all these compounds a large number of bands are observed at higher energy ranges. Nevertheless, one can identify four major band systems for $[CrOCl_4]^-$ and $[CrOCl_5]^{2-}$ species in the region 11000–13000, 18000–19000, 22000–24000 and 25000–40000 cm^{-1}. In the lone example of $[CrOF_5]^{2-}$, the first absorption band appears at considerably lower energy, 8300 cm^{-1}.

The interpretation of the two major lower energy bands in the oxyhalo complexes has been a subject of controversy; the issue is not yet settled. For a considerable period these two bands were considered to be due to $d_{xy} \rightarrow d_{xz,yz}$ ($^2B_2 \rightarrow {}^2E$) and $d_{xy} \rightarrow d_{x^2-y^2}$ ($^2B_2 \rightarrow {}^2B_1$) transitions. While $^2B_2 \rightarrow {}^2E$ transition appears to be alright, however, the notion of $^2B_2 \rightarrow {}^2B_1$ transition has been questioned on the basis of theoretical calculations[180–183] as well as experimental observations. It may be noted in Table 6 that the intensity of absorption occurring in solution at ca. 18000 cm^{-1} is an order of magnitude stronger than the one observed at ca. 13000 cm^{-1}. Since $^2B_2 \rightarrow {}^2B_1$ transition is symmetry forbidden, so it should not be more intense than the symmetry allowed $^2B_2 \rightarrow {}^2B_1$ transition. The same reasoning applies also to the polarized spectra of $[Ph_4As][CrOCl_4]$.

SCF–MO calculations of $[CrOF_5]^{2-}$ [180] have provided transition energies that are in good agreement with the observed spectra. Thus the absorption occurring at 8300 cm^{-1} has been related to $O_{2p_\pi} \rightarrow d_{xy}$ (7800 cm^{-1}) and $d_{xy} \rightarrow d_{yz,zx}$ (9100 cm^{-1}) transitions. The second absorption band at ca. 15000 cm^{-1} has been related to $F_{2p_\pi} \rightarrow d_{xy}$ (16700 cm^{-1}). The calculated energies for $d_{xy} \rightarrow d_{x^2-y^2}$ and $d_{xy} \rightarrow d_{z^2}$ are 24400 and 33400 cm^{-1}, respectively, which may be related to some of the weak absorptions occurring in these regions.

The features observed in the polarized spectra of $[Ph_4As][CrOCl_4]$ have been interpreted with the aid of CI calculations[181]. The lowest energy bands at ca. 13000 and 18000 cm^{-1} are assigned to $d_{xy} \rightarrow d_{yz,zx}$ and Cr–O(π) → Cr–O(σ*) transitions, respec-

Table 6. Electronic spectral features of chromium(V) oxyhalo complexes

Compound	Medium	Temperature, K	ν_{max}, 10^3 cm^{-1} (ε, M^{-1} cm^{-1})	Ref.
$[Hpy][CrOCl_4]$	Nujol mull	298	13.3, 18.5, 23.0, 25.8, 31.5, 39.3, 43.1	166
$[Hquin][CrOCl_4]$	Nujol mull	298	13.1, 17.8, ca. 23.5, ca. 27(sh), 32.8, 45.2,	166
	CH_2Cl_2	298	13.2(14), 18.2(160)	166
$[Hi\text{-}quin][CrOCl_4]$	Nujol mull	298	13.1, 18.1, 22.5, 24.8br, 30.1	166
$[Me_4N][CrOCl_4]$	Nujol mull	298	12.4, 18.1, 21.0(sh), ca. 23, ca. 26.5, ca. 37(br)	166
	Nujol mull	77	11.9, 12.5, 13.2, 18.6, 22.5	169
	MeCN	298	12.5(18), 18.0(80), 25.2, 26.1, 26.9, 27.6(1570)	169

Table 6 (continued)

Compound	Medium	Temperature, K	ν_{max}, 10^3 cm^{-1} (ε, M^{-1} cm^{-1})	Ref.
$[Et_4N][CrOCl_4]$	Nujol mull	298	13.2, 18.0, 22br(sh), 26.3, 31.0, 37.3	166
	Nujol mull	77	12.4, 13.3, 14, 14.7, 18.3, 22.1	169
	CH_2Cl_2	298	13.2(19), 18.1(129), 22.5(1600), 24.7(2400), ca. 26(2200)	166
	CH_2Cl_2	298	13.1(22), 18.1(167), 24.9, 41.8	169
	MeCN	298	12.5(17), 17.9(99), 24.4(sh), 25.5(sh), 26(sh), 26.7, 27.3	169
$[PrN][CrOCl_4]$	Nujol mull	77	11.5, 12.1, 12.6, 13.0, 13.3, 17.2(sh), 18.3, 22.7	169
	MeCN	298	12.7(15), 18(77), 22.6(362)	169
	CH_2Cl_2	298	13.5(21), 18.1(148), 22.9(sh), 25.1(1780), 26(sh), 26.8(sh), 27.6(sh), 32.2(2015)	166
$[Ph_4As][CrOCl_4]$	Crystal	4	12.9, 13.4(sh), 13.8, 14.4, 15.3, 17.8, 18.2, 18.5, 18.7, 20.3, 22, 23(sh), 24.5	181
	Nujol mull	77	13, 13.8, 14.4, 18.4, 23.1	169
	Nujol mull	298	13.4, 18.1, 21(sh), 23(sh), 25(sh)	166
	CH_2Cl_2	298	13.0(19), 18.1(129), 23(sh, 1650), 25.7(sh, 2200), 36.8(8600), 37.8(10 000), 38.8(10 000)	166
	MeCN	298	12.4(21), 18(80), 23(sh), 25.0(770), 25.9, 26.9, 27.6, 28.5, 37.1, 37.9(4740), 38.9, 39.9	169
$[BzPh_3P][CrOCl_4]$	$MeNO_2$	77	12.0, 12.7, 13.7, 17.9, 22.2	169
	MeCN	298	12.5, 18.0(74), 23.1(sh), 25.3, 26.1, 26.9, 27.7(910), 28.5, 29.3, 36.7, 37.5(5000), 38.5	169
$Rb_2[CrOCl_5]$	Nujol mull	77	11.9, 12.9, 13.6, 14.5, 15.0, 17.3(sh), 21.4, 24.2	169
$Cs_2[CrOCl_5]$	Nujol mull	77	10.8, 11.6, 12.4, 18.5	169
	MeCN	298	12.4(19), 18.0(76), 22.8(sh), 25.2, 26.1, 27.0, 27.7(7550), 28.5, 29.2, 40.5(14 500)	169
	Nujol mull	298	11.4, 19.1, ca. 20.2(sh), 22.8, 31.2, ca. 37.7(br), ca. 45.5	166
$[H_2(2,2')\text{-bipy}][CrOCl_5]$	Nujol mull	298	12.1, 19.2(sh), 23.3(sh), 24.8, 34.4, 37.5, 44.9	166
$[(2,2')\text{-bipy}CrOCl_3]$	Nujol mull	298	15.0, 18.6, 24.2	171
$[H_2(4,4')\text{-bipy}][CrOCl_5]$	Nujol mull	298	11.8, ca. 19(sh), 24.0, ca. 28(br), 37.2, 43.8	166
$[Et_4N]_2[CrOF_5]$	Nujol mull	77	8.3, 8.4, 14.9, 15.5, 22.3, 25.5, 26.2, 27.0, 27.7, 28.6, 29.4, 30.2, 31.1, 36.8(sh)	169
	CH_2Cl_2	298	16.5(0.8), 22.2(17), 29.1(105), 35.8(sh), 36.0(446), 39.4(324)	169
	MeCN	298	14.1, 14.8, 15.6(0.8), 22.2(18), 22.7(18), 27.3(sh), 28(sh), 28.6(sh), 29.4, 30.1(sh), 30.8(sh), 31.5(sh), 36.9(118), 38.6, 39.6, 40.7, 41.7	169

tively. Several intense absorptions observed below 20000 cm^{-1} are ascribed to O → Cr and Cl → Cr charge transfer transitions.

MO calculations of $[CrOX_4]^-$ using SCF–MS–X_α method have been independently carried out by two groups[182, 183]. In one case[182] calculation of the excited states suggested that the lowest energy absorption is mainly due to $d_{xy} \rightarrow d_{yz,zx}$ transition with some possible additional features due to halogen to metal charge transfer excitations. The second absorption and other higher energy bands are all predicted[182] to arise from halogen to metal charge transfer which obscure the symmetry forbidden $d_{xy} \rightarrow d_{x^2-y^2}$ excitation. Indeed, the calculated $d_{xy} \rightarrow d_{x^2-y^2}$ transition energies[183] for $[CrOCl_4]^-$ (19980 cm^{-1}), $[CrOCl_5]^{2-}$ (20740 cm^{-1}) and $[CrOF_5]^{2-}$ (24700 cm^{-1}) species are in good agreement with the observed band positions.

Similar to electronic spectra, EPR spectra of chromium(V) oxyhalo complexes have been widely studied[169, 172, 175–179, 183–188]. It should be noted, however, that many of the earlier investigations were carried out with materials that were chemically impure. Selected EPR data for some of these compounds are given in Table 7.

Table 7. EPR spectral parameters of $[CrOXn]^{m-}$ complexes

Compound	$g_{\parallel}$	$g_{\perp}$	⟨g⟩	⟨A⟩ (10^{-4} cm^{-1})	Ref.
[Hpy][$CrOCl_4$]	2.000	1.978	1.985		177, 178
[Hquin][$CrOCl_4$]			1.991		177
[Me_4N][$CrOCl_4$]			1.989		177
			1.990		169
[Et_4N][$CrOCl_4$]			1.987		169
[Ph_4As][$CrOCl_4$]	2.006	1.979	1.988	18.1	185
$[CrOF_4]^-$	1.959	1.968	1.964	21.4	179
$[NH_4]_2[CrOCl_5]$	2.008	1.977		18.2	184
$K_2[CrOCl_5]$	2.008	1.977			178
$Rb_2[CrOCl_5]$			1.986		169
$Cs_2[CrOCl_5]$			1.988		169
$[CrOF_5]^{2-}$	1.959	1.968	1.963	23.1	179
$[Et_4N]_2[CrOF_5]$			1.968		169

An interesting difference in the EPR spectral features of the oxychloro complexes from the oxofluoro species can be noted in Table 7. For $[CrOCl_n]^{m-}$, $g_{\parallel} > g_{\perp}$, but a reversal of the trend occurs in fluoro complexes. To rationalize this observation few suggestions have been made. In one explanation[179] difference in spin-orbit interaction of the ligand(X) is held responsible for the observed reversal. More recently it has been argued[183] that the rather high energy of the $^2B_1$ excited state arising from occupied b_1 orbital of fluoro complexes makes significant difference in EPR spectra of these species with respect to the chloro analogs.

It has been shown from change in EPR spectra that the addition of an excess halide ion to a solution of $[CrOX_4]^-$ ion leads to the formation of corresponding $[CrOX_5]^{2-}$

species[179, 185]. In several cases equilibrium constants for the reaction

$$CrOCl_4^- + L \overset{K}{\rightleftharpoons} CrOCl_4L^-$$

have been determined[185]. Few such values of K are given here

L = Ph_3PO, K = 3 ± 2; L = $(Me_2N)_3PO$, K = 463 ± 130; L = py, K = 160; L = Cl(Ph_4AsCl), K = 43 ± 10; L = Cl(Et_4NCl), K = 250 ± 33.

6 Alkoxides, Amides and Alkyls

Synthesis and characterization of the alkoxides, amides and alkyls of chromium(IV) mark significant advancement in the chemistry of this oxidation state. In contrast, except an alkoxo complex with perfluoropinacol, other types of compounds are still unkown for pentavalent chromium.

6.1 Alkoxides

At least six tertiary alkoxides of chromium(IV) viz. $Cr(OBu^t)_4$, $Cr(OBu^t)_2(OCMe_2Et_2)_2$, $Cr(OCMe_2Et)_4$, $Cr(OCMeEt_2)_4$, $Cr(OCEt_3)_4$ and $Cr(OSiEt_3)_4$ have been characterized[189–193]. Two secondary alkoxides $Cr(OCHMeBu^t)_4$[194] and $Cr(OCHBu^t_2)_4$[195] are also known. However, no such compounds have been obtained with primary alcohols. The key to the success of preparing chromium(IV) alkoxides is to use alcohols that would exert maximum steric crowding so that the metal centre is least amenable to electron-transfer reactions.

The first example of the formation of chromium(IV) alkoxide was provided by $Cr(OBu^t)_4$ which was obtained[189] by heating dibenzene chromium(0) in benzene with di-t-butylperoxide in a sealed tube at 90 °C. Subsequently this compound has been prepared by several other methods:

(i) By reacting Bu^tOH with $Cr(NEt_2)_4$[190].
(ii) By the oxidation of $Cr(OBu^t)_3$ in presence of Bu^tOH with various reagents such as $Bu^t_2O_2$, $CrO_2(OBu^t)_2$, $Pb(OAc)_4$, Br_2 and O_2[190].
(iii) By reacting $CrCl_3 \cdot 3\,THF$ (THF = tetrahydrofuran) with $NaOBu^t$ and an oxidising agent such as CuCl, $Bu^t_2O_2$, $CrOCl_3$ and $Cr(OCOPh)_2Cl_3$[191].
(iv) By the oxidation of $LiCr(OBu^t)_4$ with CuCl in THF[192].

Other tertiary alkoxides have been prepared by alcoholysis of $Cr(NEt_2)_4$[190, 192, 193] according to the reaction

$$Cr(NEt_2)_4 + 4\,ROH \rightarrow Cr(OR)_4 + 4\,Et_2NH$$

$Cr(OBu^t)_4$ is a solid and melts at 36–37 °C, but other derivatives are liquid at room temperature. They are all deep blue in color and have monomeric composition. The room temperature magnetic moments of the tertiary alkoxides (ca. 2.8 B.M.) are close to the spin-only value[193]. They are further characterized by a ν(Cr–O) stretch at ca. 625 cm^{-1}[193] in the IR spectra. $Cr(OBu^t)_4$ has considerable thermodynamic stability as evident from its high heat of formation, $\Delta H_f = -1275$ kJ mol^{-1}[194]. In general all tertiary alkoxides are air and moisture sensitive, although $Cr(OBu^t)_4$ is less susceptible to hydrolysis than $Ti(OBu^t)_4$. Taking advantage of the slow rate of alcohol exchange for $Cr(OBu^t)_4$, a mixed alkoxide $Cr(OBu^t)_2(OCMe_2Et)_2$, has been isolated[193]. Chromium(IV) tertiary alkoxides are generally reduced by primary and secondary alcohols, e.g.

$$2\,Cr(OBu^t)_4 + 7\,RCH_2OH \rightarrow 2\,Cr(OCH_2R)_3 + RCHO + 8\,Bu^tOH$$

A similar reaction with acetylacetone (Hacac) yields $Cr(acac)_3$.

Although primary and secondary alcohols get oxidized by $Cr(OBu^t)_4$ to the corresponding aldehydes or ketones, yet the first stable secondary alkoxide of chromium(IV), $Cr(OCHMeBu^t)_4$, was obtained by heating a mixture of $Cr(OBu^t)_4$ and pinacolylalcohol in a sealed tube[195]. The other known secondary alkoxide, $Cr(OCHBu^t_2)_4$ has been prepared[196] either by reacting $Li(OCHBu^t_2)$ with $CrCl_3$ in diethylether (the product formed plausibly by disproportionation reaction) or by oxidizing $LiCr(OCHBu^t_2)_4$ with traces of oxygen or CuCl. Both of the secondary alkoxides are highly air and moisture sensitive. The room temperature magnetic moment of $Cr(OCHMeBu^t)_4$ is 3.1 B.M.[195].

The alkoxides of chromium(IV) do not give EPR signal in solution even at 100 K[193, 195]. At 10 K a broad absorption occurs for $Cr(OBu^t)_4$ at $g \approx 4$ which is considered to be due to the forbidden $\Delta Ms = \pm 2$ transition and a relatively sharp signal at $g = 1.962$ due to $\Delta Ms = \pm 1$ transition[193].

The electronic spectra of chromium(IV) tertiary alkoxides show four absorption bands[193]. For example, in $Cr(OBu^t)_4$, these are observed at 9100 (ε, 10), 15200 (ε, 560), 25000 (ε, ca. 500) and 41000 (ε, 10000) cm^{-1}[193]. These spectra have been interpreted on the basis of a local tetrahedral symmetry with $^3A_2$ as the ground state. The lowest energy band, which is very weak, is due to symmetry forbidden $^3A_2 \rightarrow {}^3T_2(F)$ transition; fairly intense next two bands arise due to excitations to $^3T_1(F)$ and $^3T_1(P)$ states, respectively. According to these assignments for $Cr(OBu^t)_4$, 10 Dq = 9430 cm^{-1} and B = 795 cm^{-1}[193].

$Cr(OCHBu^t_2)_4$ is the only chromium(IV) alkoxide whose X-ray crystal structure has been determined[196]. This compound crystallizes in monoclinic form with the space group *C2/c*. The O–Cr–O angles are close to tetrahedral (108.8, 110.3, 112.4°), but show slight tetragonal flattening in the direction of the two-fold axis. The Cr–O bond lengths (av. 1.77 Å) are very short which probably would indicate a considerable degree of O–Cr p_π-d_π bonding. The rather large Cr–O–C angles (140.5, 141.1°) also support this view[196].

Salts of composition $M[CrO(pfp)_2]$ (M = K, Cs, Et_4N) obtained with perfluoropinacol (H_2pfp), $(CF_3)_2C(OH)(OH)C(CF_3)_2$, represent the only known alkoxo compounds of chromium(V). It is remarkable that although primary and secondary alcohols get readily oxidized by acidified chromates(VI), yet the compounds $M[CrO(pfp)_2]$ have considerable stability towards reduction and hydrolysis. Deep blue crystals of the potassium and caesium salts are readily obtained[197, 198] by dissolving K_2CrO_4 or Cs_2CrO_4 and H_2pfp (1:2 molar ratio) in ethanol-water mixture acidified with H_2SO_4 and maintaining the solution at the boiling point for ca. 10 min. The Et_4N salt can be obtained by metathesis[198]. The room temperature magnetic moments of the potassium and caesium salts have expected values (1.71 B.M.)[197], but the moment of Et_4N salt (2.33 B.M.)[198] is rather high. The optical absorption spectrum of $K[CrO(pfp)_2]$ consists of four bands at ca. 17000, 26000, 28500 and 32000 cm^{-1}; the last two bands show some solvent dependency. The EPR spectrum of this compound in DMF at 77 K shows rhombic symmetry: $g_x = 1.985$, $g_y = 1.977$, $g_z = 1.974$; $A_x = 14.8$ G and $A_z = 35.1$ G[198].

6.2 Amides

Four stable chromium(IV) dialkylamides, $Cr(NR_2)_4$ (R = Et, Pr^n, Bu^n: piperidide), have been characterized[190, 199]. Initially $Cr(NEt_2)_4$ was obtained[190] in poor yield in a reaction involving $LiNEt_2$ and $CrCl_3$. Subsequently all $Cr(NR_2)_4$ compounds have been prepared in high yield by thermal disproportionation of $Cr(NR_2)_3$ *in vacuo*[199]. The reaction evidently moves in the forword direction due to much greater

$$[Cr(NR_2)_3]_2 \rightarrow Cr(NR_2)_4 + [Cr(NR_2)_2]_n$$

volatility of $Cr(R_2N)_4$ with respect to dimeric $Cr(NR_2)_3$ and polymeric $Cr(NR_2)_2$. All $Cr(NR_2)_4$ compounds are green liquids, but Cr(piperidide)$_4$ is a green solid melting at 60 °C. The magnetic susceptibilities of these compounds obey Curie-Weiss law, and the room temperature magnetic moments are ca. 2.8 B.M.[199]. Chromium(IV) tetraamides react with primary and secondary alcohols to yield chromium(III) alkoxides and aldehydes or ketones. However, as already mentioned, the reaction with tertiary alcohol or trialkylsilanol provides an easy mean to prepare tetraalkoxide and tetrakis-trialkylsiloxide of chromium(IV)[193]. $Cr(NEt_2)_4$ differs from other tetravalent metal amides, $M(NEt_2)_4$ (M = Ti, V, Zr), in its reactivity towards CS_2. As against the formation of normal tetrakis-dithiocarbamates, in the case of chromium, $Cr(S_2CNEt_2)_3$ is formed[199]. The reaction with CO_2 takes place in a more complex fashion: the products $Cr_2^{III}(O_2C\text{-}NEt_2)_4(\mu\text{-}NEt_2)_2$ and $Cr_2^{II}(O_2CNEt_2)_4 \cdot 2\,HNEt_2$ have been structurally characterized[200].

The electronic absorption spectrum of $Cr(NEt_2)_4$ gives a single d–d band at 13700 cm^{-1} (ε, 1200) and several intense absorptions in the range 25000–50000 cm^{-1} (ε, $\sim 10^4$–10^5) probably due to charge transfer and intraligand transitions[199]. Judging the high intensity of absorption, the band at 13700 cm^{-1} cannot be ascribed to the $^3A_2 \rightarrow {}^3T_2(F)$ transition in Td symmetry. It could be considered as the allowed $^3A_2 \rightarrow {}^3T_1(F)$ transition, but that would yield a value of 10 Dq $\sim$ 7000 cm^{-1} which is significantly low compared to that of $Cr(OBu^t)_4$ (9430 cm^{-1}), and also would require the transition $^3A_2 \rightarrow {}^3T_1(P)$ to appear at ca. 21000 cm^{-1}. Since the 10 Dq value in isoelectronic tetrahedral vanadium(IV) is greater with amides than alkoxides, it has been suggested[197, 199] that $Cr(NEt_2)_4$ probably distorts from T_d to D_{2d} symmetry. In D_{2d} symmetry the T_2 states are split into an orbital singlet and a doublet, $B_2 + E$; similarly T_1 into $A_2 + E$. The ground state becomes $^3B_1$ and transitions are allowed to $^3A_2$ and ^3E. This would mean that the absorption occuring at 13700 cm^{-1} could be due to either $^3B_1 \rightarrow {}^3A_2$[199] or $^3B_1 \rightarrow {}^3E$[201] transition.

6.3 Alkyls

As many as twelve σ-bonded chromium(IV) tetraalkyls have been isolated in pure form. These include seven alkane derivatives, viz. $Cr(CH_2SiMe_3)_4$, $Cr(CH_2CMe_3)_4$,

$Cr(CH_2CMe_2Ph)_4$, $Cr(CH_2CPh_3)_4$, $CrMe_4$, $CrBu^t_4$ and $Cr(Pr^i)_4$; four bridged alicyclic systems, viz. $Cr(nor)_4$(I), $Cr(1\text{-}cam)_4$(II), $Cr(1\text{-}adme)_4$(III) and $Cr(2\text{-}adm)_4$(IV); and one alkenyl derivative, $Cr(CPh{=}CMe_2)_4$.

The major problem associated with the generation of transition metal alkyl compounds is that they undergo thermal decomposition if processes such as α- or β-elimination of a metal hydride, homolysis, or coupling of the ligands at the metal center can occur. The most prevalent reaction is hydride elimination

$$MCH_2CH_2R \rightarrow MH + CH_2 = CHR$$

Stabilization of chromium(IV) tetraalkyls have been accomplished by using alkyl groups that do not contain a β-hydrogen atom, and by conferring a tetrahedral environment to the metal center through steric control.

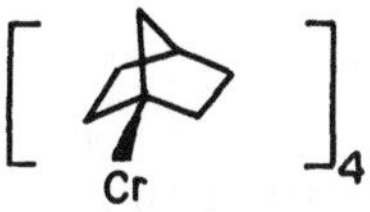

$Cr(nor)_4$
nor = norbornyl
I

$Cr(1\text{-}cam)_4$
1-cam = 1-camphoryl
II

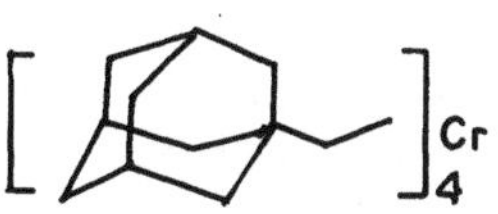

$Cr(1\text{-}adme)_4$
1-adme = 1-adamentamethyl
III

$Cr(2\text{-}adm)_4$
2-adm = 2-adamentyl
IV

Essentially three different chemical routes are adopted for preparing organochromium(IV) compounds. These are: (i) controlled aerial oxidation of $[CrR_4]^-$, which is generated by treating $CrCl_3 \cdot 3\,THF$ with lithium or magnesium alkyl halide in ether or hydrocarbon solvents; (ii) thermal or photochemical disproportionation of chromium(III) trialkyls; (iii) exchange reaction involving $Cr(OBu^t)_4$ and lithium alkyls. Preparative methods of individual chromium(IV) tetraalkyls and their characteristics[201–209] are summarized in Table 8. Aside from the compounds shown in Table 8, few more tetraalkyls have been reported which are either thermally unstable or did not give satisfactory elemental analyses. For example, $Cr(PhCH_2)_4$[201] was obtained as a red liquid and $Cr(1\text{-}adm)_4$[208] as a red solid in impure form. $CrBu^n_4$ and $CrBu^i_4$ formed in solution[205], but could not be isolated. Reactions of Grignard reagents with $Cr(OBu^t)_4$ has been reported[210] to produce mixed tetraalkyl intermediates

$$Cr(OBu^t)_4 + nRMgX \rightarrow R_nCr(OBu^t)_{4-n} + nBu^tOMgX$$

that readily decompose to trialkyl (or triaryl) chromium(III).

Table 8. Preparative methods and characteristics of chromium(IV) tetraalkyls

Compound	Preparative method	Characteristics	Ref.
$Cr(CH_2SiMe_3)_4$	$[Cr(CH_2SiMe_3)_4]^-$ formed by adding a solution of $Mg(CH_2SiMe_3)Cl$ in Et_2O to $CrCl_3 \cdot 3$ THF is oxidized to $Cr(CH_2SiMe_3)_4$ by the addition of a saturated aqueous solution of NH_4Cl.	Purple red crystals; m.p., 40 °C; μ_{eff} 2.9 B.M.; ν_{max} 17100, 19400 (ε, 1060) cm^{-1}; air sensitive; thermally stable; does not react with most organic solvents, CS_2, phosphines, amines etc.	202, 203
$Cr(CH_2CMe_3)_4$ $Cr(CH_2CMe_2Ph)_4$ $Cr(CH_2CPh_3)_4$	Grignard or lithium reagent derived from respective alkyl halide is treated with $CrCl_3$. 3 THF; chromium(IV) tetraalkyl forms either through disproportionation or controlled aerial oxidation of the reaction product.	$Cr(CH_2CMe_3)_4$: maroon crystals; m.p. 110 °C; μ_{eff} 2.7 B.M.; ν_{max} 18500, 21100 (ε, 1090) cm^{-1}. $Cr(CH_2CMe_2Ph)_4$: purple prisms; m.p., 120 °C (decomp); μ_{eff} 2.8 B.M.: ν_{max} 18200, 20500 (ε, 1380) cm^{-1}. $Cr(CH_2CPh_3)_4$: purple prisms; m.p., 130 °C (decomp); μ_{eff} 2.6 B.M.; ν_{max} 17600, 20200 (ε, 1380) cm^{-1}.	201, 204, 205
$CrMe_4$ $CrBu^t_4$	By reacting lithium alkyl with $Cr(OBu^t)_4$ in pentane.	$CrMe_4$: maroon oil; m.p., ca. − 60 °C; thermally unstable; ν_{max} 20000, 22000 (ε, 600) cm^{-1}. $CrBu^t_4$: dark red solid; μ_{eff} 2.6 B.M.; ν_{max} 15630(sh), 20620(850) cm^{-1}.	203, 205
$CrPr^i_4$	U.V. irradiation of the product formed by reacting $MgPr^iBr$ with $CrCl_3$.	Red liquid: m.p., ca. 20 °C; μ_{eff} 2.8 B.M.	206
$Cr(nor)_4$(I) $Cr(1\text{-cam})_4$(II)	By reacting $CrCl_3 \cdot 3$ THF with lithium alkyl in pentane.	Red-brown solid: air stable compounds: stable to attack by dil. H_2SO_4; μ_{eff} 2.84(I), 2.96(II) B.M.	207
$Cr(1\text{-adme})_4$(III)	By reacting Li(1-adme) with $Cr(OBu^t)_4$ in light petroleum.	Maroon solid: decomposes at 164 °C without melting; inert to water, but solutions are air sensitive; ν_{max} 20000 (ε, 1000) cm^{-1}.	208
$Cr(2\text{-adm})_4$(IV)	By reacting $Cr(OBu^t)_4$ with $Mg(2\text{-adm})_2$ in light petroleum.	Maroon solid: decomposes at 125 °C without melting; air stable in the solid state.	208
$Cr(CPh{=}CMe_2)_4$	$CrCl_3 \cdot 3$ THF is reacted with $LiCPh{=}CMe_2$ in Et_2O at − 78 °C.	Green crystals: μ_{eff} 2.81 B.M.; ^1H NMR, 1.36(CMe_2), ca. 7.18(Ph) p.p.m.; reacts slowly with oxygen in the solid state.	209

Polarographic studies of several chromium(IV) tetraalkyls have shown[201] that they reversibly get reduced to chromium(III) tetraalkyls. Coluometric measurements[201] confirmed one-electron transfer in the redox system. The $E_{1/2}$ values (Table 9) obtained in ethanol solution using Bu_4NI as the supporting electrolyte indicated that the reduction becomes more difficult as the alkyl chains contain more phenyl rings. This means that the stability of chromium(IV) tetraalkyls decreases with the depletion of aromatic rings. Indeed this conclusion receives support from a comparative study of the reactivity of chromium(IV) tetraalkyls toward oxygen which show[201] the following trend: $CrMe_4 > Cr(CH_2SiMe_3)_4 \sim Cr(CH_2CMe_3)_4 \gg Cr(CH_2CMe_2Ph)_4 \gg Cr(CH_2CPh_3)_4$. The above

Table 9. Polarographic half-wave potentials for chromium(IV) tetraalkyls

Compound	$E_{1/2}$(V vs. SCE)
$Cr(CH_2SiMe_3)_4$	−1.28
$Cr(CH_2CMe_3)_4$	−1.65
$Cr(CH_2CMe_2Ph)_4$	−1.97
$Cr(CH_2CPh_3)_4$	−1.99

decreased order of reactivity indicates that with augmentation of steric crowding the metal atom becomes less accessible to attack by oxygen.

The electronic spectra of several chromium(IV) tetraalkyls show[201] a very weak absorption at ca. 15 000 cm^{-1} and a strong doublet centering at ca. 20 000 cm^{-1}. As noted earlier if T_d symmetry distorts to D_{2d} symmetry, then the allowed transition $^3A_2 \rightarrow {}^3T_1$ of the former (T_d) will split into two components $^3B_1 \rightarrow {}^3E$ and $^3B_1 \rightarrow {}^3A_2$ in the latter (D_{2d}). Similarly the forbidden $^3A_2 \rightarrow {}^3T_2$ transition of T_d can split into two components, $^3B_1 \rightarrow {}^3E$ and $^3B_1 \rightarrow {}^1A_1$, in D2d. The weak band at 15 000 cm^{-1} has been considered[201] due to $^3B_1 \rightarrow {}^3E$ transition, and the strong doublet at 20 000 cm^{-1} due to the symmetry allowed components. On this basis, the 10 Dq value has been estimated[201] to be ca. 14 500 cm^{-1} and B ca. 450 cm^{-1}.

X-ray crystal structure has been determined for two tetraalkyl chromium(IV) compounds, $Cr(CH_2CMe_2Ph)_4$[211] and $Cr(CPh{=}CMe_2)_4$[209, 212]. $Cr(CH_2CMe_2Ph)_4$ crystallizes in monoclinic form with *P2₁/c* space group. The Cr–C distances (2.07, 2.06, 2.05, 2.01 Å) and C–Cr–C angles (114, 112, 111, 107, 106, 105°) show distortion from ideal tetrahedral geometry and confirms flattened tetrahedral configuration suggested from electronic spectra[201]. Similar distortion from ideal tetrahedral[201] configuration has also been observed with $Cr(CPh{=}CMe_2)_4$ (monoclinic, space group $P2_1/c$) in which the average Cr–C bond length is 2.036 Å, two C–Cr–C angles are 116.2° and two others are 106.2[212].

The EPR spectra of several chromium(IV) tetraalkyls have been investigated in considerable detail [201, 208, 212–214]. In general these compounds, in solution, at room temperature give a broad signal near $g \simeq 2$ and have features expected for a triplet species

Table 10. Zero-field splitting parameters of chromium(IV) tetraalkyls

Compound	T (K)	D (cm^{-1})	E (cm^{-1})	Ref.
$CrMe_4$	110	0.005	ca. 0	201
$Cr(CH_2CMe_3)_4$	145	0.011	ca. 0	201
$Cr(CH_2CMe_2Ph)_4$	110	0.021 0.049	ca. 0	201
$Cr(CH_2CPh_3)_4$	110	0.07	0.01	201
$Cr(CH_2SiMe_3)_4$	110	0.073 0.089	ca. 0	201
$Cr(1\text{-nor})_4$	93	0.027 0.023 0.013 0.012	0.0041 0.0027 0.0032 0.0029	214

(S = 1) with a random rotational motion. This signal shifts slightly at lower temperature and an additional weaker signal appears at $g \simeq 4$. The low-field signal is assigned to the nominally forbidden $\Delta Ms = \pm 2$ transition. In the cases of $Cr(CH_2CMe_2Ph)_4$ and $Cr(CH_2SiMe_3)_4$, at still lower temperature, unexpected line broadening occur which is interpreted due to the presence of two S = 1 species[201]. Single crystal EPR spectrum of $Cr(nor)_4$ shows isotropic feature (g = 1.99) at room temperature[214]. However, $Cr(nor)_4$ apparently exists in four conformations in isooctane and cyclohexane glasses that are distinguishable on the basis of their zero-field splitting parameters. Table 10 lists zero-field splitting parameters of some of the chromium(IV) tetraalkyls.

7 Tertiary Hydroxy Carboxylates

Despite the fact that chromium(IV) tertiaryalkoxides are quite stable compounds, no such compounds have been obtained with tertiary hydroxy carboxylic acids ($R_3C(OH)CO_2H$). On the other hand, a series of remarkably stable, water soluble chromium(V) complexes of the general formula $Na[OCr(O_2COCR_1R_2)_2]$ (Va-g) have been isolated recently[215].

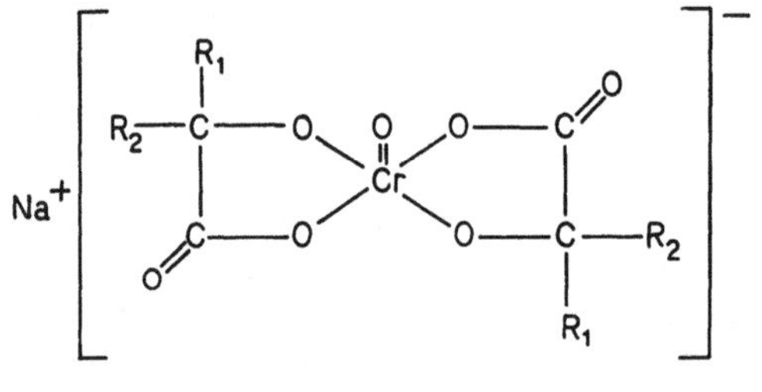

V
a: R_1 = Me, R_2 = Et
b: $R_1 = R_2$ = Me
c: $R_1 = R_2$ = Et
d: $R_1 = R_2$ = Bu^n
e: R_1, R_2 = $(CH_2)_4$
f: R_1, R_2 = $(CH_2)_5$
g: R_1 = Me, R_2 = Ph

These compounds are obtained in high yield by reacting anhydrous $Na_2Cr_2O_7$ in acetone with tertiary α-hydroxy acids, viz. 2-methyl-2-hydroxypropionic, 2-methyl-2-hydroxybutyric, 2-ethyl-2-hydroxybutyric, 2-butyl-2-hydroxyhexanoic, 1-hydroxycyclopentanecarboxylic, 1-hydroxycyclohexanecarboxylic and 2-phenyl-2-hydroxypropionic acids[215]. The stoichiometry of the reaction is

$$Na_2Cr_2O_7 + 5\,R_1R_2C(OH)CO_2H \rightarrow 2\,Na[OCr(O_2COCR_1R_2)_2] + R_1R_2CO + CO_2 + 5\,H_2O$$

The first member of the series (Va) was initially obtained[216] as $K[OCr(O_2CCOMeEt)_2] \cdot H_2O$ from a reaction involving CrO_3 and 2-methyl-2-hydroxybutyric acid in aqueous solution. As opposed to the reaction in acetone here the reaction stoichiometry is

$$2\,HCrO_4^- + 4\,Et(Me)C(OH)CO_2H + 4\,H^+ \rightarrow [OCr(EtMeCOCO_2)_2]^- + Cr^{3+} + 2\,MeCOEt + 2\,CO_2 + 7\,H_2O$$

The isolation of the product was quite tedious in this case as it required the use of ion-exchange removal of Cr^{3+} ion. The formation of unstable chromium(V) complexes with various primary and secondary hydroxy acids (e.g. glycolic, glyoxalic, tartaric, malic, mandellic), dicarboxylic acids (e.g. oxalic, malonic, methylmalonic) and tricarboxylic acids such as citric acid take place in solution[215], but the product cannot be isolated in solid state. There has been extensive kinetic studies on chromic acid oxidation of various organic substrates including many carboxylic acids, and the rate laws conform to the formation of chromium(IV) and chromium(V) intermediates. This aspect, however, will not be considered here in detail; interested readers will find references[217–222] as useful sources of information. Suffice it would be to say here that the chromium(V) complexes are formed[215] either as a sequel to two consecutive one-electron reactions

$$Cr(VI) + S \rightarrow Cr(V) + R$$
$$Cr(VI) + R \rightarrow Cr(V) + P$$

or through an initial two-electron reduction of chromium(VI) to chromium(IV), which in turn gets oxidized to chromium(V)

$$Cr(VI) + S \rightarrow Cr(IV) + P$$
$$Cr(VI) + Cr(IV) \rightarrow 2\,Cr(V)$$

Concerning the stabilities of chromium(V) hydroxy carboxylates, higher alkyl groups tend to stabilize the complexes, but phenyl groups have strong destabilizing effect[215]. They are more soluble and more stable in acetone than in water. However, their decomposition in aqueous solution can be retarded by adding small amount of the corresponding free hydroxy acid. This suggests that the decomposition in aqueous solution, which is predominantly disproportionation of chromium(V) to chromium(VI) and chromium(III), is preceeded by hydrolysis.

Chromium(V) hydroxy carboxylates are characterized by the presence of a sharp ν(Cr=O) vibration in the IR spectra at ca. 1000 cm^{-1}. In the uv/visible region several absorption bands are observed, which typically for (Va) appear at 12500 (ε, 35), 13300 (ε, 41), 13640 (ε, 42), 15480 (ε, 34), 19600 (ε, 170), 20370 (ε, 164), 28560 (ε, 1210), 40000 (ε, 6420) cm^{-1} [216]. The room temperature magnetic moment of (Va) is 2.05 B.M., and the EPR spectrum of the aqueous solution gives a sharp signal at $g = 1.978$ along with a four-line hyperfine structure due to ^{53}Cr ($I = 3/2$)[216].

X-ray crystal structure of $K[OCr(O_2CCOMeEt)_2] \cdot H_2O$ has been determined[216]. The compound crystallizes in monoclinic form with the space group *Cc*. The anion geometry is intermediate between square pyramidal and trigonal bipyramidal. The three types of Cr–O distance are: Cr=O, 1.554 Å; av. Cr–O (hydroxy), 1.781 Å; av. Cr–O (carboxy), 1.911 Å.

8 Macrocyclic Compounds

In the past few years several interesting N_4-macrocyclic compounds (see structures VI–IX for the abbreviations used for the ligands) of chromium(IV) and chromium(V) have been discovered. These include compounds of the types: $[Cr^{IV}O(phthalocyanine)]_2$, $[Cr^{IV}O(porphyrins)]$, $[Cr^{V}O(corrole)]$, $[Cr^{V}OCl(porphyrins)]$ and $[Cr^{V}N(porphyrins)]$.

The first evidence for the formation of $[Cr^{IV}O(OEP)]$ (H_2OEP = octaethylporphyrin) in solution was obtained[223] from the cyclic voltammetric study of Cr(OEP)(OH) in BuCN. The one-electron oxidation that occured at 0.79 V was considered due to the

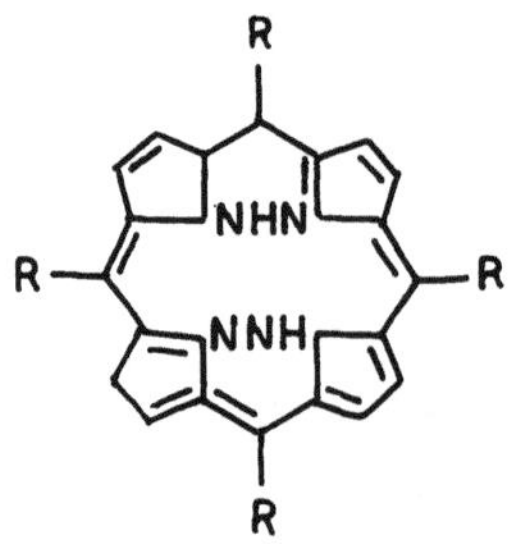

H_2TPP: $R = C_6H_5$
H_2TTP: $R = p - MeC_6H_4$
H_2TMP: $R = 2,4,6\text{-}Me_3C_6H_2$
VI

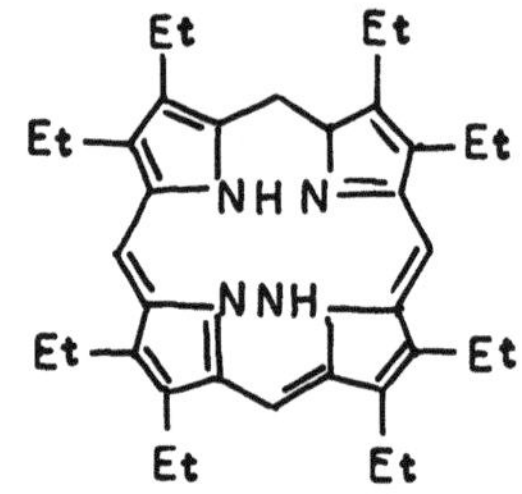

H_2OEP
VII

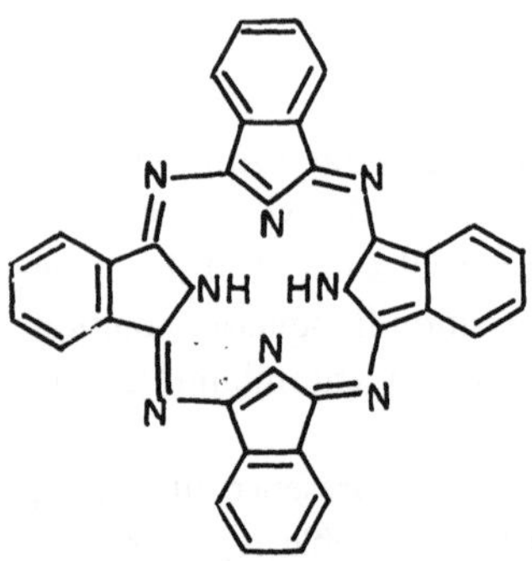

H_2Pc
VIII

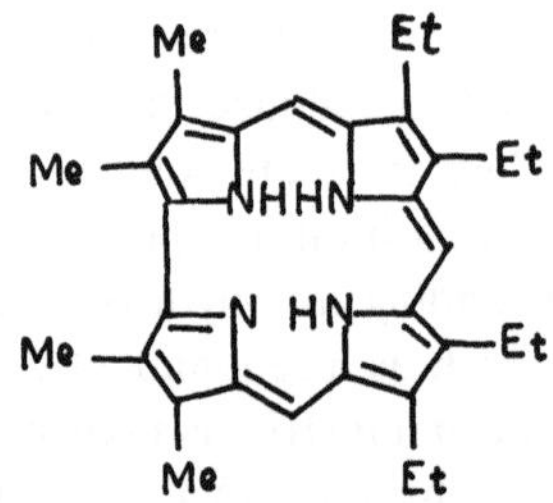

H_3MEC
IX

formation of a chromium(IV) species which, however, was too unstable to be characterized spectroscopically. The first stable chromium(IV) macrocyclic compound [CrO(Pc)] (H_2Pc = phthalocyanine) has been prepared by aerial oxidation of β-Cr^{II}(Pc)[224]. From IR, Raman and mass spectra of the blue-violet product obtained by treating the purple β-Cr(Pc) with $^{16}O_2$, $^{16}O^{18}O$ and $^{18}O_2$, [CrO(Pc)] has been characterized[224] to be a dimeric compound. The magnetic moments of this compound at 300 K (1.9 B.M.) and 77 K (1.6 B.M.) also indicate strong magnetic interaction due to dimerization. Similar to Cr^{II}(Pc), the reaction of dioxygen with Cr^{II}(TPP) (H_2TPP = tetraphenylporphyrin) leads to the formation of [Cr^{IV}O(TPP)][225]. Almost simultaneously, two groups of workers[226, 227] reported the preparation of CrO(TPP) and several other chromium(IV) porphyrinates, viz. CrO(TPP), CrO(OEP) and CrO(TMP). The preparative methods consist either oxidation of [Cr^{III}(OH)(TTP)] · 2 H_2O and [Cr^{III}(OH)(OEP)] · 0.5 H_2O with sodium hypochlorite[226], or by reacting Cr^{III}(TPP)Cl, Cr^{III}(TTP)Cl and Cr^{III}(TMP)Cl with iodosyl benzene[227]. Other oxidizing agents, such as, *t*-butylhydroperoxide and *m*-chloroperoxybenzoic acid have also been used[227]. $Cr^{(IV)}$O(porphyrinates) are all bright-red diamagnetic solids, and show an intense band in the IR spectra at ca. 1020 cm^{-1} (νCr=O) which shifts to 980 cm^{-1} on ^{18}O substitution. The *o*-phenyl protons of CrO(TTP) appear as two sharp doublets at − 20 °C in the ^1H NMR spectra and become broad singlets at 25 °C, and finally coalesce at 37 °C. From variable temperature ^1H NMR data the rotational barrier for ring rotation has been calculated[227] to be 15.7 kcal mol^{-1}

in toluene $- d_8$. Unlike FeO(TTP) or MnO(TTP), CrO(TTP) does not hydroxylates alkanes or converts alkenes to epoxides, and therefore does not serve as a model for cytochrome P450 enzymes. It does oxidize, however, benzyl alcohol showing peroxidase II behavior[226)]. PPh_3 abstracts oxygen from Cr^{IV}O(porphyrins) to yield Cr^{II}(porphyrins) and PPh_3O[227)]. X-ray crystal structures have been determined for CrO(TPP)[225)] and CrO(TTP)[227)]. CrO(TPP) crystals are tetragonal (space group, I_4/m), but CrO(TTP) crystallizes in monoclinic form with space group $P2_1/c$. The description of CrO(TPP) structure is: the chromium atom is displaced from the plane of the porphyrin ring towards oxygen atom, and the four pyrrole nitrogen atoms are in the plane of the porphyrin. The Cr–O distance is 1.62 Å and the avarage Cr–N distance is 2.036 Å[225)]. In CrO(TTP) the Cr–O distance is 1.572 Å and the avarage Cr–N distance is 2.032 Å[227)]. However, in contrast to CrO(TPP), the porphyrine ring in CrO(TTP) is saddle shaped with the pyrrole β-carbons displaced 0.340 and 0.586 Å above and below the mean pyrrole nitrogen plane. In this compound the chromium atom is 0.469 Å above the avarage pyrrole nitrogen plane. Based on the fact that the structural analysis of CrO(TTP) was carried out in greater detail and there was some disordering problem with CrO(TPP), perhaps it may be commented that in CrO(TPP) also the porphyrin ring is non-coplanar.

Cr^{V}O(TPP)Cl is formed as an unstable product on treating a CH_2Cl_2 solution of Cr^{III}(TPP)Cl with excess of PhIO at room temperature[228)]. This compound decomposes on standing to the previously described Cr^{IV}O(TPP). The EPR spectra of the chromium(V) complex, its ^{17}O derivative and its Bu^tNH_2 adduct show strong hyperfine interaction with the metal nucleus and the metal ligands. Thus, in CrO(TPP)Cl: $\langle A \rangle_{Cr}$ = 23 G, $\langle a \rangle_{N(TPP)}$ = 2.85 G; in CrO^{17}(TPP)Cl: $\langle a \rangle_{^{17}O}$ = 5.4 G[228)]. CrO(TPP)Cl has been treated as a model for cytochrome P-450[229)]. Facile epoxidation of alkenes occur by PhIO in presence of Cr(TPP)Cl as catalyst

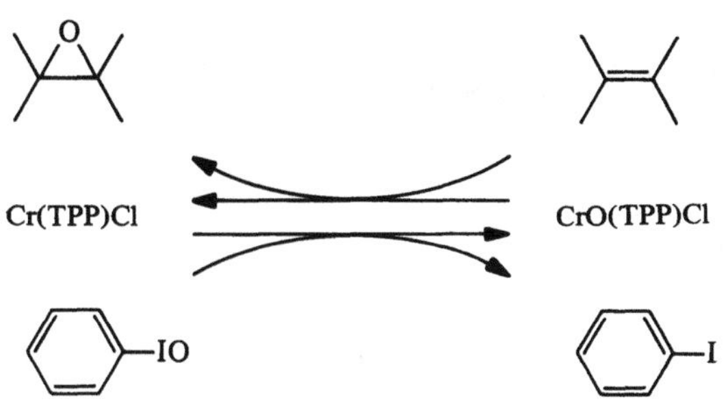

A stable oxochromium(V) complex CrO(MEC) with 2,3,17,18-tetramethyl-7,8,12,13-tetraethylcorrole (H_3MEC, IX) has been prepared[230)] by reacting $CrCl_2$ with H_3MEC and NaOAc in boiling dimethylformamide. The isotropic EPR spectrum gives $\langle g \rangle = 1.987$, $\langle A \rangle_{Cr} = 19.3 \times 10^{-4}\ cm^{-1}$ and $\langle a \rangle_N = 3.3 \times 10^{-4}\ cm^{-1}$. This compound has been further characterized from its IR and XPS spectra[230)].

Very recently two independent studies have reported[231, 232)] the synthesis of remarkably stable nitrido(porphyrinato)chromium(V) complexes, CrN(OEP), CrN(TTP) and CrN(TPP). In one method[231)] [Cr^{III}(OH)(TTP)] · 2 H_2O and [Cr^{III}(OH)(OEP)] · 0.5 H_2O are oxidized with sodium hypochlorite in presence of ammonia, while in the other method[232)] a dichloromethane or benzene solution of [Cr^{III}(TPP/TTP)N_3] is subjected to photo irradiation. All three compounds have been characterized by uv/visible, IR, EPR and mass spectral studies[231, 232]. The IR spectra show the presence of $\nu Cr{\equiv}N$ in $Cr^{V}N(TTP)$ at 1017 cm^{-1} which shifts to 991 cm^{-1} on ^{15}N substitution[232]. At room temperature, CrN(TTP) in solution, gives isotropic spectrum with $\langle g \rangle = 1.982$, $\langle A \rangle_{^{53}Cr} = 28.3$ G and $\langle a \rangle_{^{14}N} = 2.7$ G. The 11-line superhyperfine splitting shows the magnetic equivalence of four pyrrole nitrogens and the axial nitrido ligand. In frozen glass at 77 K the spectrum shows axial symmetry: $g_{\parallel} = 1.958$, $g_{\perp} = 1.994$; $A_{\perp ^{53}Cr} = 22.4$ G, $A_{\parallel ^{53}Cr} = 40.1$ G[231, 232]. The axial and equatorial superhyperfine splitting constants have been determined from ENDOR spectra at 3.6 K[231]. The EPR and ENDOR parameters indicate that there is strong spin localization in the d_{xy} orbital, but the d_{z^2} orbital is practically free from spin localization. X-ray crystal structure determined for CrN(TPP) · C_6H_6 (monoclinic, space group *P2₁/c*)[232] shows that the Cr–N(nitrido) distance, 1.565 Å, is very short and consistent with formal chromium-nitrogen triple bond, and the avarage Cr–N(pyrrole) distance is 2.04 Å. Again, as observed with CrO(TTP) (but not reported for CrO(TPP)) the porphyrine ring is nonplanar in CrN(TPP); it has excursions 0.56 Å below and 0.29 Å above the plane of the pyrrole nitrogens. CrN(TTP) is reduced by PPh_3 and $P(OMe)_3$, the reaction products are $[Cr^{III}(Ph_3P{=}N)(TTP)]$ and $[Cr^{III}(NMePO(OMe)_2)(TTP)]$[233].

The redox behavior of a few macrocyclic systems have been investigated. The redox potentials (vs. SCE) obtained from cyclic voltammetric measurements are given below.

$$Cr^{III}(OEP)(OH) \underset{BuCN/Bu_4N[ClO_4]}{\overset{E_{1/2} = 0.79\ V}{\rightleftharpoons}} Cr^{IV}O(OEP) \qquad 233$$

$$\underset{(H_2PP = \text{protoporphyrin})}{Cr^{IV}O(PP)} \underset{CH_2Cl_2/Bu_4N(ClO_4)}{\overset{E_{1/2} = 0.91\ V}{\rightleftharpoons}} [Cr^{V}O(PP)]^{+} \qquad 234$$

$$[Cr^{VI}(MEC)]^{3+} \xleftarrow{E_{irrv} = 0.63\ V} Cr^{V}O(MEC) \underset{CH_2Cl_2/Bu_4N(ClO_4)}{\overset{E_{1/2} = 0.33\ V}{\rightleftharpoons}} Cr^{III}(MEC) \qquad 235$$

9 Other Types of Compounds

Chromium compounds in higher oxidation states (> 3) have been obtained[236a, 236b] with *cis*-1,2-disubstituted ethylene-1,2-dithiolato group $R_2C_2S_2$ in which R = CF_3, CN, Ph. These complexes have the general composition $[Cr(S_2C_2R_2)_3]^n$: R = CF_3, n = 0, −1, −2; R = CN, n = 0, −1; R = Ph, n = 0, −1. It should be noted, however, that 1,2-dithiolenes being unorthodox ligands, the oxidation state formalism is difficult to apply in these complexes.

Chromium complexes of bis(trifluoromethyl)-1,2-dithiene are the most well-characterized species[236a]. $[Cr(S_2C_2(CF_3)_2)_3]$ has been obtained by reacting $Cr(CO)_6$ with $(CF_3)_2C_2S_2$ in cyclohexane. Reduction of this compound with hydrazine followed by the addition of Ph_4AsCl affords $[Ph_4As]_2[Cr(S_2C_2(CF_3)_2)_3]$. Again, oxidation of $[Ph_4As]_2[Cr(S_2C_2(CF_3)_2)_3]$ with one-equivalent of $[Cr(S_2C_2(CF_3)_2)_3]$ results in the formation of $[Ph_4As][Cr(C_2S_2(CF_3)_2)_3]$. When the trivalent chromium complex $[Ph_4As]_3[Cr(S_2C_2(CN)_2)_3]$ is oxidized by $[Mo(S_2C_2(CF_3)_2)_3]$ the product has the composition $[Ph_4As][Cr(S_2C_2(CN)_2)_3]$[236a]. $[Cr(S_2C_2Ph_2)_3]^n$ (n = 0, − 1) complexes have also been obtained[236b] in pure state.

$[Cr(S_2C_2(CF_3)_2)_3]$ is diamagnetic, while the room temperature magnetic moments of the anionic compounds[236a] are: $[Ph_4As][Cr(S_2C_2(CF_3)_2)_3]$, 1.89 B.M.; $[Ph_4As]_2[Cr(S_2C_2(CF_3)_2)_3]$, 2.95 B.M.; $[Ph_4As]_2[Cr(S_2C_2(CN)_2)_3]$, 2.89 B.M. These values should indicate $d^0(Cr^{VI})$, $d^1(Cr^V)$, $d^2(Cr^{IV})$ configuration of the metal ions for n = 0, − 1, − 2, respectively. However, such simple attribution of the oxidation states to the metal atom cannot be rationalized with the EPR results. The EPR data of n = − 1 species are

$[Cr(S_2C_2(CF_3)_2)_3]^{-1}$ [236a]: $\langle g \rangle = 1.994$; $\langle A \rangle(^{53}Cr) = 12$ G

$g_{\parallel} = g_{\perp} = 1.995$ (in glassy state)

$[Cr(S_2C_2Ph_2)_3]^{-1}$ [236b]: $\langle g \rangle = 1.996$; $\langle A \rangle = 19$ G

Analysis of the EPR data assuming D_3 symmetry of the complex has been shown[236a] to be inconsistent with d^1 configuration of the metal ion in a trigonal field.

The preparative reactions of the complexes can be considered as redox reactions in which the metal is oxidized to the formally + 4 to + 6 states, and the ligand is reduced to the thietene radical anion or dianion. The interrelationship of the complexes (n = 0, − 1, − 2) through sequential one-electron transfers can be appreciated by considering their polarographic half-wave potentials ($E_{1/2}$)[236a] measured in acetonitrile solutions

$$[Cr(S_2C_2(CF_3)_2)_3]^0 + e^- \underset{}{\overset{+1.14\ V}{\rightleftharpoons}} [Cr(S_2C_2(CF_3)_2)_3]^{-1}$$

$$[Cr(S_2C_2(CF_3)_2)_3]^{-1} + e^- \overset{+0.76\ V}{\rightleftharpoons} [Cr(S_2C_2(CF_3)_2)_3]^{-2}$$

$$[Cr(S_2C_2(CF_3)_2)_3]^{-2} + e^- \longrightarrow \text{reduction does not occur up to } -1.5\ V\,.$$

$$[Cr(S_2C_2(CN)_2)_3]^{-1} + e^- \overset{+0.16\ V}{\rightleftharpoons} [Cr(S_2C_2(CN)_2)_3]^{-2}$$

The potential date show that $[Cr(S_2C_2(CF_3)_2)_3]$ is a strong oxidizing agent; it is reduced more easily than the corresponding molybdenum and tungsten complexes. Also it may be noted that the complex containing strongly electron – withdrawing cyano group is more stable compared to the trifluoromethyl analogue.

In order to account for magnetic susceptibilities and EPR results the following electronic configuration for the complexes has been suggested[236a]

n = 0, $d^2(S = 0)$
n = − 1, $d^2l^1(S = 1/2)$
n = − 2, $d^3l^1(S = 1)$

where l^1 represents a molecular orbital localized primarily on the ligands. On the basis of this description one can consider the compounds $Cr(S_2C_2R_2)_3$ (R = CF_3, Ph) as chromium(IV) complexes, and their one-electron reduction products are also formally tetravalent chromium species, but the added electron is residing on ligand orbitals.

Another compound $(\eta^5\text{-}C_5Me_5)_2Cr_2S_2$[236c] which contains an unusual η^1(-μ-disulfide) ligand may also probably be considered as a formally chromium(IV) complex. This compound has been obtained by reacting $[(\eta^5\text{-}C_5Me_5)(CO)_2Cr]_2$ with excess of sulfur in toluene. The compound is diamagnetic and the ^1H NMR spectrum gives only one singlet at δ 2.13 ppm indicating a symmetric structure with respect to the C_5Me_5 groups. The formal oxidation state of chromium in this compound is not obvious, however, taking C_5Me_5 ligands as monoanions and the three different types of sulfur ligands as dianions, each of the chromium then acquires + 4 oxidation state. The diamagnetic behavior of this compound can be rationalized by taking into consideration a Cr–Cr double bond. The black-green crystals of $[(\eta^5\text{-}C_5Me_5)Cr_2S_5]$ are monoclinic and have space group $P2_1/c$[236c]. The important feature of the structure (X) is that the plane of the five sulfur atoms is perpendicular to the metal-metal bond and parallel to the two $\eta^5\text{-}C_5Me_5$ planes. The three different types of sulfur ligands are: (i) μ-S ligand, S(3), bridging the two chromium atoms in the usual way; (ii) a η^2(μ-S, μ-S) ligand S(1)–S(2) forming a side-one bonded disulfur bridge; (iii) a η^1(μ-S, S) ligand in which S(4) is coordinated to both the metal atoms, but S(5) remains free. The Cr–Cr distance in this molecule is 2.489 Å and the avarage Cr–S distances are: Cr–S(1) = 2.298, Cr–S(2) = 2.295, Cr–S(3) = 2.239, Cr–S(4) = 2.349 Å. The S(1)–S(2) and S(4)–S(5) distances are 2.149 and 2.101 Å, respectively.

An unique hydrido chromium(IV) compound, $[CrH_4(dmpe)_2]$ (dmpe = dimethylphosphino ethane), has been reported[237] recently, which incidentally is the only phosphino complex of either tetra- or pentavalent chromium, and also provides the first example of a eight coordinate chromium(IV) compound. This compound has been prepared by reacting $LiBu^n$ with $CrCl_2(dmpe)_2$ under hydrogen atmosphere. Apparently chromium(II) got oxidized to chromium(IV) and hydrogen reduced to the hydride ion. $[CrH_4(dmpe)_2]$ is a yellow solid and is diamagnetic. The hydride protons appear as a quintet and the proton-coupled ^{31}P NMR signal also gives quintet multiplicity. ^1H NMR: δ 1.23 ppm (PMe_2 and PCH_2, singlet), − 6.91 ppm (Cr–H, quintet; $J_{P\text{-}H}$ = 56.1 Hz); $^{31}P\{^1H\}$NMR: δ 78.8 ppm. Even at − 80 °C, no change in NMR line shape occurs which indicates that the molecule remains fluxional at this temperature. This compound is further characterized by its Cr–H stretching vibrations at 1757, 1725 and 1701 cm^{-1}. X-ray crystal structure of the compound (monoclinic, space group $P2_1/c$) has been determined[237]. Avarage Cr–P and Cr–H distances are 2.255 and 1.57 Å, respectively. The structure (XI) shows an approximate dodecahedral geometry.

Similar to porphyrins the nitrido complexes of chromium(V) have also been reported with other types of ligands. In fact the possibility of generation of chromium(V) nitrido complexes by photo-irradiation of chromium(II) aminopolycarboxylates containing coordinated azide ion were reported earlier[238] than Cr^VN (porphyrinates). Thus, photolysis of the aqueous solutions of $[Cr(edta)(N_3)]^{2-}$ (edta = ethylenediaminetetraacetate ion), $[Cr(nta)(N_3)(H_2O)]^-$ (nta = nitrilotriacetate ion) and $[Cr(vda)(N_3)_2]$ (vda = D-valine-N,N-diacetate ion) at room temperature give practically identical EPR spectra that are characteristic of chromium(V) species. For example, the EPR spectra of the nta complex at room temperature is isotropic with $\langle g \rangle = 1.969$, and of axial symmetry at 77 K, $g_{\parallel}$ = 1.932 and $g_{\perp}$ = 1.987[238]. Although these photo products could not be isolated in solid state, nevertheless they had sufficient stability in solution to be separated by ion-exchange chromatography and identified by absorption and CD spectra[238]. The same workers[239], however, have been successful to isolate the complex $[Cr^VN(salen)]$ (H_2salen = bis-salicylaldehydeethylenediamine) by irradiating $[Cr^{III}(salen)(N_3)(H_2O]$ at 313 nm. The product has been characterized to be an authentic chromium(V) species. Another chromium(V) complex with the same ligand, $[Cr^VO(salen)][PF_6]$ has been prepared[240]

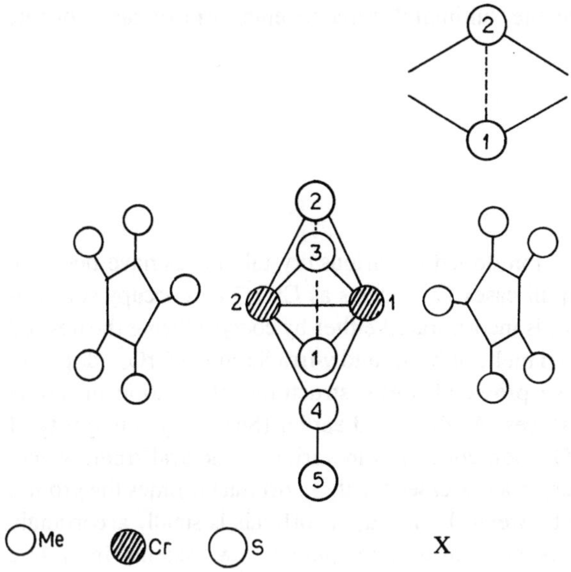

X

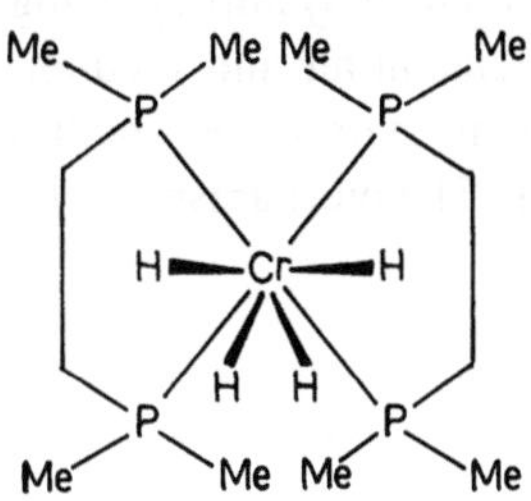

XI

by reacting $[Cr^{III}(salen)(H_2O)_2][PF_6]$ with PhIO in acetonitrile solution. Compounds containing other anions (F, Cl, Br) have also been obtained. X-ray crystal structure determined for $[Cr^{V}O(salen)][PF_6]$[240] shows disordering in the Cr=O sites on either sides of the salen plane. The avarage metal ligand bond distances are: Cr=O, 1.56 Å; Cr–O, 1.85 Å; Cr–N, 2.00 Å. The out-of-plane displacement of chromium is 0.52 Å. Similar to $Cr^{III}(TPP)Cl$, $[Cr^{III}(salen)(H_2O)_2]^+$ catalyzes epoxidation of a large number of alkenes with PhIO[240].

A compound of composition $Cr^{V}(phen)_2(IO_6)$ (phen = 1,10-phenanthroline) has been obtained[241] by treating an aqueous solution of cis-$[Cr^{III}(phen)_2(H_2O)_2](NO_3)_3 \cdot 2\,H_2O$ with $NaIO_4$ at pH ~ 4. The room temperature magnetic moment of this compound is 1.94 B.M. and its EPR spectrum shows axial symmetry.

Finally mention may be made of the observation[242] that incubation of chromate(VI) with rat liver microsomes and NaDPH produces an EPR signal characteristic of chromium(V). This along with the finding[243] that relatively stable chromium(V) species are rapidly produced when chromate(VI) is reacted with ribonucleotides, but not deoxyribonucleotides. Their persistence in vitro for a relatively long period probably suggests[243] that chromium(V) may be the "ultimate" carcenogenic form of carcenogenic chromium compounds.

10 Doped Species

EPR spectral characteristics of Cr^{5+} ion doped in various crystal lattices have been the subject of considerable interest. In all cases Cr^{5+} exists as CrO_4^{3-} and occupy substitutional lattice sites. The CrO_4^{3-} center is incorporated either by cocrystallizing the desired compound and the host material from melt, or by X- and γ-irradiation of CrO_4^{2-} doped in crystals. Detailed EPR studies have provided useful structural information including phase transition behavior of host lattices. As discussed earlier (Sect. 2.2) in majority of cases the unpaired electron of CrO_4^{3-} ion contained in various structural frame works occupy the d_{z^2} ground state, however, in a few cases the $d_{x^2-y^2}$ orbital becomes the ground state. Since the separation energy between d_{z^2} and $d_{x^2-y^2}$ orbitals is small, accordingly spin-lattice relaxation times have been measured in quite a few systems from line broadening measurements at different temperatures. It is interesting to note that polymerization of ethylene by Phillips process uses a catalyst that is obtained by impregnating aqueous solution of CrO_3 on silica followed by thermal treatments in air and vacuum. EPR studies have shown[244-246] that chromium in the pentavalent state in tetrahedral coordination is the active catalytic agent in Cr_2O_3/SiO_2 catalysts. The mechanism of the polymerization reaction is believed to be

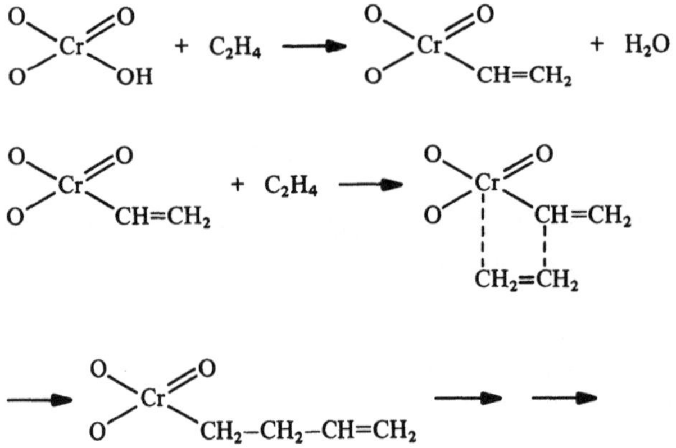

Compared to doped Cr^{5+} ion little is known about substitutionally occupied Cr^{4+} ion. Table 11 summarizes pertinent information for Cr^{4+} and Cr^{5+} ions implanted in various crystal lattices.

Table 11. EPR spectral characteristics of Cr^{4+} and Cr^{5+} ions in host lattices

Ionic species	Lattice	Comments	Ref.
Cr^{4+}	α-Al_2O_3	$g_{\parallel} = 1.90$; $g_{\parallel} = 1.913$, $g_{\perp} = 1.80$; $D = 7.55\ cm^{-1}$; $E = 0.030\ cm^{-1}$. $Dq = 2150\ cm^{-1}$, $B = 660\ cm^{-1}$, $C = 3200\ cm^{-1}$	247–249
	Si	$\langle g \rangle = 1.9962$, $\langle A \rangle = 12.54$ G. The four ^{53}Cr lines are further splitted by neighbouring ^{29}Si nuclei.	250
Cr^{5+} (dz^2 ground state)	Ca_2PO_4Cl	$g_{\parallel} = 1.9936$, $g_{\perp} = 1.9498$: $A_z = 7.6$, $A_y = 19.1$, $A_x = 22.7 \times 10^{-4}\ cm^{-1}$. In this structure CrO_4^{3-} tetrahedron is squashed along the z-axis bisecting the two OCrO (119.1°) angles.	79
	$Ca_5(PO_4)_3Cl$	Principal g- and A-values are: $g_x = 1.9329$, $g_y = 1.9466$, $g_z = 1.9790$; $A_x \geq 27.6$, $A_y \leq 21.9$, $A_z \leq 5.5 \times 10^{-4}\ cm^{-1}$. At low temperature the hexagonal structure *(P6₃/m)* changes to a centrosymmetric monoclinic phase *($P2_1/b$)*.	85–87
	$Ca_5(PO_4)_3F$	$g_x = 1.9494$, $g_y = 1.9489$, $g_z = 1.9774$; $A_x = 27.8$, $A_y = 21.2$, $A_z = 4.7 \times 10^{-4}\ cm^{-1}$. The room temperature hexagonal phase undergoes small structural change at low temperature to a phase which has noncentrosymmetric space group.	86–87
	$Ca_5(PO_4)_3OH$	The EPR spectra of the doped crystals grown from $CaCl_2$ melt have the following principal g- and A-values: $g_x = 1.9479$, $g_y = 1.9489$, $g_z = 1.9768$; $A_y = 20.4 \times 10^{-4}\ cm^{-1}$. The hexagonal structure is stabilized by chloride impurities down to 4.2 K.	87
	$Sr_5(PO_4)_3Cl$	$g_x = 1.9348$, $g_y = 1.9430$, $g_z = 1.9822$; $A_x = 26.5$, $A_y = 22.1$, $A_z = 6.7 \times 10^{-4}\ cm^{-1}$. No phase transformation of the hexagonal structure occurs up to 4.2 K.	88
	$Ba_5(PO_4)_3Cl$	$g_x = 1.9222$, $g_y = 1.9317$, $g_z = 1.9737$; $A_x = 26.2$, $A_y = 22.3$, $A_z = 6.5 \times 10^{-4}\ cm^{-1}$. The hexagonal phase is retained even at 4.2 K. A comparison of the results of Ca-, Sr- and Ba apatites shows that the cation substitution has discernible effect on the PO_4^{3-} structure.	89
	YPO_4	$g_{\parallel} = 1.9772$, $g_{\perp} = 1.9525$; $A_{\perp} = 8.5$, $A_{\parallel} = 19.6 \times 10^{-4}\ cm^{-1}$.	90
	YVO_4	$g_{\parallel} = 1.9773$, $g_{\perp} = 1.9511$; $A_{\parallel} = 8.6$, $A_{\perp} = 18.7 \times 10^{-4}\ cm^{-1}$. The PO_4 and VO_4 tetrahedra are distorted to D_{2d} symmetry by a c-axis stretch and although $d_{x^2-y^2}$ ground state is expected, but d_{z^2} is observed. This inversion is attributed to strong covalency. However, the g- and A-values cannot be explained with D_{2d} symmetry, and the width of superhyperfine lines is orientation and magnetic field dependent. These anomalies are probably due to motional effects.	90
	Li_3PO_4	$g_x = 1.7411$, $g_y = 1.7722$, $g_z = 1.9402$; $A_x = 30$, $A_y \geq 19$, $A_z = 11.5 \times 10^{-4}\ cm^{-1}$.	91
	Li_3VO_4	$g_x = 1.9146$, $g_y = 1.8980$, $g_z = 1.9398$; $A_x = 21.9$, $A_y = 25.1$, $A_z = 4.5 \times 10^{-4}\ cm^{-1}$.	91
	Li_3AsO_4	$g_x = 1.9383$, $g_y = 1.9364$, $g_z = 1.9734$; $A_x = 22.3$, $A_y = 24.5$, $A_z = 2 \times 10^{-4}\ cm^{-1}$. In all three Li_3XO_4 systems unusual g-values are observed and the deviation is very pronounced with Li_3PO_4. The unusual g-values are interpreted to be due to very small distortion of the CrO_4^{3-}	91

Table 11 (continued)

Ionic species	Lattice	Comments	Ref.
		tetrahedron from the ideal Td symmetry and the corresponding small splitting of the ground state doublets.	
	Be_2SiO_4	$g_x = 1.951$, $g_y = 1.958$, $g_z = 1.993$; $\langle A \rangle = 25.2$ G	251
	$CaMoO_4$	$g_\parallel = 1.992$, $g_\perp = 1.945$; $A_\parallel = 3.6$, $A_\perp = 25.7$ G. Spin-lattice relaxation time and phase-memory time have been determined.	252
	$CaWO_4$	$g_\parallel = 1.988$, $g_\perp = 1.943$; $A_\parallel = 4$, $A_\perp = 23 \times 10^{-4}\ cm^{-1}$.	253, 254
	K_2SO_4	$g_x = 1.926$, $g_y = 1.954$, $g_z = 1.928$. EPR and ENDOR studies indicate that CrO_4^{3-} centers interact with a nearby proton formed during X-irradiation.	256
	$(NH_4)_2SO_4$	At 300 K: $g_x = 1.983$, $g_y = 1.986$, $g_z = 1.946$. At 208 K: $g_x = 1.971$, $g_y = 1.988$, $g_z = 1.946$. The soft-mode phase transition occurs at 223 K.	257
	K_2CrO_4	$g_x = 1.9776$, $g_y = 1.9539$, $g_z = 1.9281$. CrO_4^{3-} is generated by γ-irradiation.	258, 259
	MoO_3	Two inequivalent Cr^{5+} sites have been located. I: g_x, $g_y = 1.975$, $g_z = 1.944$; II: $g_x = 1.969$, $g_y = 1.958$, $g_z = 1.824$.	260
	$SrTiO_3$	The EPR spectrum of Cr-doped $SrTiO_3$ annealed at 1000 K in an oxygen atmosphere is attributed to Cr^{5+} substitutional to Ti^{4+} in tetragonal symmetry. $g_\parallel = 1.961$, $g_\perp = 1.920$: $A_\parallel = 34.7$, $A_\perp = 10.4 \times 10^{-4}\ cm^{-1}$. Below 105 K splitting of the EPR lines occur which is due to tetragonal to orthorhombic phase transition. In the orthorhombic phase: $g_x = 1.981$, $g_y = 1.956$, $g_z = 1.945$; $A_x < 3$, $A_y = 16$, $A_z = 28 \times 10^{-4}\ cm^{-1}$.	261
		Nonlinear Jahn-Teller coupling and local dynamics of $SrTiO_3 : Cr^{5+}$ near the structural phase transition has been investigated.	262, 263
	CaV_2O_6	At 300 K: $g_x = 1.953$, $g_y = 1.984$, $g_z = 1.933$.	264
	BaV_2O_6	At 300 K: $g_x = 1.981$, $g_y = 1.983$, $g_z = 1.901$. At 77 K: $g_x = 1.980$, $g_y = 1.984$, $g_z = 1.900$. The point symmetry of CrO_4^{3-} depends on temperature. At 77 K the Cr^{5+} ion is probably displaced to one of the two neighbouring V^{5+} ions.	264
	KCl	$g_x = 1.948$, $g_y = 1.953$, $g_z = 1.967$.	265
	Al_2O_3	Isotropic spectrum $\langle g \rangle \simeq 1.97$ with a line width ~ 50 G.	266, 267
Cr^{5+} ($d_{x^2-y^2}$ ground state)	$NaVO_3$	$g_\parallel = 1.897$, $g_\perp = 1.983$: $A_\parallel = 36.6$, $A_\perp = 6.9 \times 10^{-4}\ cm^{-1}$. $g_\parallel < g_\perp$ suggests that the ground state is $d_{x^2-y^2}$ orbital.	94
	SiO_2	$g_\parallel = 1.898$, $g_\perp = 1.975$. The spectrum changed significantly when contacted with H_2O, HCl and NH_3 vapor at room temperature. H_2O: $g_\parallel = 1.952$, $g_\perp = 1.972$; HCl: $g_\parallel = 2.002$, $g_\perp = 1.970$; NH_3: $g_\parallel = 1.925$, $g_\perp = 1.992$.	245
	KH_2AsO_4	$g_x = 1.983$, $g_y = 1.977$, $g_z = 1.946$.	95–99
	KH_2PO_4	$g_\parallel$ ($g_x \approx g_y$) $= 1.98$, $g_\perp$ (g_z) $= 1.95$.	
	$(NH_4)_2HAsO_4$	$g_\perp$ (g_z) $= 1.955$, $g_\parallel$ ($g_x \approx g_y$) $= 1.974$. The $d_{x^2-y^2}$ ground state is observed despite the fact that the tetrahedron is compressed by 1.6%. This has been attributed to the presence of anchoring protons in the oxygen	

Table 11 (continued)

Ionic species	Lattice	Comments	Ref.
		planes perpendicular to the crystallographic direction along which squashing occurs. ENDOR spectrum showed that the $dx^2 - y^2$ orbital couples with two lateral protons along c-axis of the crystal.	

11 Note Added in Proof

After the submission of this article another review article dealing with coordination chemistry of chromium(V) has appeared[268]. There is considerable difference in scope and emphasis between this article and the present one. For example, the formation of chromium(V) compounds and decay in the course of chromium(VI) reduction – a topic that has been primarily left out of discussion in our article – has received detailed consideration in the other article. In many respect the two articles are complimentary to each other. In the last few months several interesting papers have come to our notice and these are annoted here. Magnetic properties of three ternary oxides of chromium(IV), viz. Sr_2CrO_4, Ba_2CrO_4 and Ba_3CrO_5 have been reported[269]. Fluoro and oxyfluoro compounds have received more attention. One paper reports[270] IR and electronic spectra of CrF_4 and CrF_5 both in the solid state and in inert gas matrices. In this paper the preparation and characterization of $Cs[CrF_6]$ and $Cs_2[CrF_6]$ are also reported. Another paper reports[271] synthesis and characterization of $[NH_4][CrF_6]$, $[NO^+][CrF_6]$ and $[NO^+]_2[CrF_6]$. Molecular structure of CrF_5 in the gas phase has been investigated[272] by electron diffraction method. Improved synthesis and properties of $CrOF_3$ and related oxo-anions, $[CrOF_4]^-$ and $[CrOF_5]^{2-}$ has been reported[273]. Electronic structure and spectroscopic properties of $[Cr(O_2)_4]^{3-}$ received further investigation[274]. There are three reports on porphyrin complexes. One of these deals with resonance Raman Spectra of the chromium(V) nitrido complex, [(TTP)CrN][275]. Several oxochromium(IV) complexes, [CrO(P)] (P = porphyrin derivatives) have been synthesized[276] by reacting $[Cr^{II}(P)]$ with molecular oxygen. [CrO(P)] react with $[Cr^{II}(P)]$ and $[Fe^{II}(P)]$ to form $[(P)Cr^{III}–O–Cr^{III}(P)]$[276] and $[(P)Cr^{III}–O–Fe^{III}(P)]$[277], respectively. Electronic spectra and photochemical behaviour of chromium(IV) tetranorbornyl has been reported[278].

12 References

1. Cotton, F. A., Wilkinson, G.: Advanced Inorganic Chemistry (4. Ed.), New York, Wiley 1980
2. Sneeden, R. P. A.: Organochromium Compounds, New York, Academic Press 1975
3. Wiede, O. F.: Chem. Ber. *30,* 2178 (1897): *31,* 516, 3139 (1898)
4. Riesenfeld, E. H., Wohlers, H. E., Kutsch, W. A.: ibid. *38,* 1885 (1905); Riesenfeld, E. H., Kutsch, W. A., Ohl, H., Wohlers, H. E.: ibid. *38,* 3380 (1905)

5. Weinland, R. F., Fridrich, W.: ibid. *38,* 3784 (1905); Weinland, R. F., Fiederer, F.: ibid. *39,* 4042 (1906); Weinland, R. F., Fiederer, F.: ibid. *40,* 2090 (1907)
6. Wöhler, F.: Gott. Nachr. 147 (1859)
7. Michel, A., Bernard, J.: Compt. Rend. *200,* 1316 (1935)
8a. Klemm, W.: Angew. Chem. *63,* 396 (1951)
8b. Scholder, R.: ibid. *65,* 240 (1953)
9. Wiberg, K. B.: Oxidation in Organic Chemistry, New York, Academic Press 1965
10. Stewart, R.: Oxidation Mechanisms Applied to Organic Chemistry, New York, Benjamin 1969
11. Espenson, J. H.: Acc. Chem. Res. *3,* 347 (1970)
12. Beattie, J. K., Haight, Jr., G. P.: Progr. Inorg. Chem. *17,* Part II, 93 (1972)
13. Kepert, D. L.: The Early Transition Metals, London, Academic Press 1972
14. Rollinson, C. L.: Pergamon Texts in Inorganic Chemistry, Vol. *21,* London, Pergamon Press 1973
15. Rosenblum, C., Holt, S. L.: in: Transition Metal Chemistry (Carlin, R. L. (ed.)), p. 87, Vol. *7,* New York, Marcel Dekker 1972
16. Wilhelmi, K.-A., Jonsson, O.: Acta Chem. Scand. *12,* 1532 (1958)
17. Kubota, B.: J. Phys. Soc. Jpn. *15,* 1706 (1960); J. Amer. Ceram. Soc. *44,* 239 (1961)
18. Thamer, B. J., Douglass, R. M., Staritzky, E.: J. Amer. Chem. Soc. *79,* 547 (1957)
19. Swoboda, T. J., Arthur, Jr., P., Cox, N. L., Ingraham, J. N., Oppegard, A. L., Sadler, M. S.: J. Appl. Phys. Suppl. 32, 374S (1961)
20. Darnell, F. J., Cloud, W. H.: Bull. Soc. Chim. France 1164 (1965)
21. Ingraham, J. N., Swoboda, T. J.: U.S. Pat. 3,034,988 (1962)
22. Arthur, P., Ingraham, J. N.: U.S. Pat. 3,117,093 (1964)
23. Dismukes, J. P., Martin, D. F., Exstrom, L., Wang, C. C., Couths, M. D.: Ind. Eng. Chem. Prod. Res. Dev. *10,* 319 (1971)
24. Montiglo, U., Aspes, P., Basse, G., Gallinotti, E.: U.S. Pat. 3,979,310 (1976)
25. Jaleel, V. A., Kannan, T. S.: Bull. Mater. Sci. *5,* 231 (1983)
26. Demazeau, G., Maestro, P., Plante, T., Pouchard, M., Hagenmuller, P.: Mater. Res. Bull. *14,* 121 (1979)
27. Demazeau, G., Maestro, P., Plante, T., Pouchard, M., Hagenmuller, P.: IEEE Trans. Magn. Mag-*16,* 9 (1980)
28. Maestro, P., Demazeau, G., Pouchard, M., Hagenmuller, P.: ibid. *18,* 1000 (1982)
29. Agarwal, D. K., Biswas, A. K., Rao, C. N. R., Subbarao, E. C.: Mater. Res. Bull. *13,* 1135 (1978)
30. DeVries, R. C.: ibid. *1,* 83 (1966)
31. Ishibashi, S., Namikawa, T., Satoun, M.: ibid. *14,* 51 (1979)
32. Cloud, W. H., Schreiber, D. S., Babcock, K. R.: J. Appl. Phys. Suppl. *33,* 1193 (1962)
33. Siratori, K., Iida, S.: J. Phys. Soc. Jpn. *15,* 210 (1960): *15,* 2362 (1960)
34. Stoffel, A. M.: J. Appl. Phys. *40,* 1238 (1969)
35. Rodbell, D. S.: J. Phys. Soc. Jpn. *21,* 1224 (1966)
36. Haneda, K., Kojima, H., Morrish, A. H., Picone, P. J., Waki, K.: J. Appl. Phys. *53,* 2686 (1982)
37. Rodbell, D. S., Lommel, J. M., DeVries, R. C.: J. Phys. Soc. Jpn. *21,* 2430 (1966)
38. Kubota, B., Hirota, E.: ibid. *16,* 345 (1961)
39. Chapin, D. S., Kafalas, J. A., Honig, J. M.: J. Phys. Chem. *69,* 1402 (1965)
40. Nakayama, N., Hirota, E., Nishikawa, T.: J. Amer. Ceram. Soc. *49,* 52 (1966)
41. Dissanayake, M. A. L., Chase, L. L.: Phys. Rev. *B18,* 6872 (1978): *B23,* 6254 (1981)
42. Goodenough, J. B.: Bull. Soc. Chim. France 1200 (1965)
43. Ikemoto, I., Kinoshita, S., Kuroda, H.: J. Solid State Chem. *17,* 425 (1976)
44. Osmolovskii, M. G., Ivanov, I. K., Kositkov, Yu. P.: Izv. Akad. Nauk SSSR, Neorg. Mater. *15,* 118 (1979)
45. Wilhelmi, K.-A.: Nature *203,* 967 (1964)
46. Wilhelmi, K.-A.: Acta Chem. Scand. *19,* 165 (1965)
47. Barker, M. G., Hooper, A. J.: J.C.S. Dalton Trans. 2487 (1975)
48. Lavielle, L., Kessler, H., Hatterer, A.: Bull. Soc. Chim. France 1918 (1973)
49. Knights, C. F., Phillips, B. A.: Chem. Soc. Spl. Pub. *30,* 134 (1976)

50. Kim, K.-Y., Johnson, G. K., O'Hare, P. A. G., Phillips, B. A.: J. Chem. Thermody. *13,* 695 (1981)
51. Scholder, R., Sperka, G.: Z. Anorg. Allgem. Chem. *285,* 49 (1956)
52. Wilhelmi, K.-A., Jonsson, O.: Acta Chem. Scand. *15,* 1415 (1961)
53. Scholder, R., Klemm, W.: Angew. Chem. *66,* 461 (1954)
54. Wilhelmi, K.-A.: Arkiv Kemi *26,* 157 (1967)
55. Goodenough, J. B., Longo, J. M., Kafalas, J. A.: Mater. Res. Bull. *3,* 471 (1968)
56. Chamberland, B. L.: Solid State Comm. *5,* 663 (1967)
57. Roth, W. L., DeVries, R. C.: J. Appl. Phys. *38,* 951 (1967); J. Amer. Ceram. Soc. *51,* 72 (1968)
58. Chamberland, B. L.: Inorg. Chem. *8,* 286 (1969)
59. Chamberland, B. L.: J. Solid State Chem. *43,* 309 (1982)
60. Chamberland, B. L.: ibid. *48,* 318 (1983)
61. Chamberland, B. L., Katz, L.: Acta Cryst. *B38,* 54 (1982)
62. Haradem, P. S., Chamberland, B. L., Katz, L.: J. Solid State Chem. *34,* 59 (1980)
63. Gai, P. L., Jacobson, A. J., Rao, C. N. R.: Inorg. Chem. *15,* 480 (1976)
64. Evans, D. M., Katz, L.: Acta Cryst. *B28,* 1219 (1972)
65. Scholder, R., Schwarz, H.: Z. Anorg. Allgem. Chem. *326,* 1 (1963)
66. Wilhelmi, K.-A., Jonsson, O.: Acta Chem. Scand. *19,* 177 (1965)
67. Scholder, R., Schwochow, F., Schwarz, H.: Z. Anorg. Allgem. Chem. *363,* 10 (1968)
68. LeFlen, G., Olazcuaga, R., Parant, J.-P., Reau, J.-M., Fouassier, C.: Compt. Rend. *C273,* 1358 (1971)
69. Olazcuaga, R., Reau, J.-M., LeFlem, G., Hagenmuller, P.: Z. Anorg. Allgem. Chem. *412,* 271 (1975)
70. Kim, K.-Y., Johnson, G. K., Johnson, C. E., Flotow, H. E., Appelman, E. H., O'Hare, P. A. G., Phillips, B. A.: J. Chem. Thermody. *13,* 333 (1981)
71. Scholder, R., Suchy, H.: Z. Anorg. Allgem. Chem. *308,* 295 (1961)
72. Scholder, R., Schwarz, H.: ibid. *326,* 11 (1963)
73. Glasser, F. P., Osborn, E. F.: J. Amer. Ceram. Soc. *41,* 358 (1958)
74. Banks, E., Jaunarajs, K. L.: Inorg. Chem. *4,* 78 (1965)
75. Klemm, W.: Angew. Chem. *63,* 396 (1951)
76. Scholder, R.: ibid. *70,* 583 (1958)
77. Johnson, W.: Miner. Mag. *32,* 408 (1960)
78. Greenblatt, M., Banks, E., Post, B.: Acta Cryst. *23,* 166 (1967)
79. Banks, E., Greenblatt, M., McGarvey, B. R.: J. Chem. Phys. *47,* 3772 (1967)
80. Schwarz, H.: Z. Anorg. Allgem. Chem. *322,* 1, 15, 129, 137 (1963): *325,* 273 (1963)
81. Roy, A., Nag, K.: J. Inorg. Nucl. Chem. *40,* 1501 (1978)
82. Manca, S. G., Baran, E. J.: J. Phys. Chem. Solids *42,* 923 (1981): J. Appl. Cryst. *15,* 102 (1982)
83. Buisson, G., Bertaut, F., Mareschal, J.: Compt. Rend. *259,* 411 (1964)
84. Roy, A., Chaudhury, M., Nag, K.: Bull. Chem. Soc. Jpn. *51,* 1243 (1978)
85. Banks, E., Greenblatt, M., McGarvey, B. R.: J. Solid State Chem. *3,* 308 (1971)
86. Greenblatt, M., Pifer, J. H., Banks, E.: J. Chem. Phys. *66,* 559 (1977)
87. Pifer, J. H., Ziemski, S., Greenblatt, M.: ibid. *78,* 7038 (1983)
88. Greenblatt, M., Kuo, J.-M., Pifer, J. H.: J. Solid State Chem. *29,* 1 (1979)
89. Forster, K., Greenblatt, M., Pifer, J. H.: ibid. *30,* 121 (1979)
90. Greenblatt, M., Pifer, J. H., McGarvey, B. R., Wanklyn, B. M.: J. Chem. Phys. *74,* 6014 (1981)
91. Greenblatt, M., Pifer, J. H.: ibid. *70,* 116 (1979)
92. Carrington, A., Ingram, D. J. E., Schonland, D., Symons, M. C. R.: J. Chem. Soc. 4710 (1956)
93. van Reijen, L. L., Cossee, P., van Haren, H. J.: J. Chem. Phys. *38,* 572 (1963)
94. Dimitrieva, V., Zonn, Z. N., Maksimov, E. V.: Sov. Phys. Solid State *15,* 405 (1973)
95. Müller, K. A., Berlinger, W.: Phys. Rev. Lett. *37,* 916 (1976)
96. Müller, K. A., Dalal, N. S., Berlinger, W.: ibid. *36,* 1504 (1976)
97. Müller, K. A., Berlinger, W.: Z. Physik. *B31,* 151 (1978)
98. Gaillard, J., Gloux, P., Müller, K. A.: Phys. Rev. Lett. *38,* 1216 (1977)

99. Dalal, N. S., Reddoch, A. H., Northcott, D. J.: Chem. Phys. Lett. *58*, 553 (1978)
100. Banks, E., Greenblatt, M., Holt, S.: J. Chem. Phys. *49*, 1431 (1968)
101. Simo, C., Banks, E., Holt, S.: Inorg. Chem. *9*, 183 (1970)
102. Milstein, J. B., Ackerman, J., Holt, S. L., McGarvey, B. R.: ibid. *11*, 1178 (1972)
103. Sipio, L. D., Oleari, L., Day, P.: J.C.S. Faraday Trans. II, *68*, 1032 (1972)
104. Borromei, R., Oleari, L., Day, P.: J. Chem. Soc. Faraday Trans. II, *75*, 401 (1979)
105. Hoffman, K. A.: Chem. Ber. *39*, 3181 (1906)
106. House, D. A., Hughes, R. G., Garner, C. S.: Inorg. Chem. *6*, 1077 (1967)
107. Amiel, J.: in: Nouveau Traite de Chimie Minerale *14* (Pascal, P., (ed.)), p. 302, Paris, Masson Co. 1959
108. Kauffman, G. B., Acero, G.: Inorganic Synthesis *8*, 132 (1966)
109. House, D. A., Garner, C. S.: Inorg. Chem. *5*, 840 (1966)
110. Ebsworth, E. A. V., Garner, C. S., House, D. A., Hughes, R. G.: Inorg. Nucl. Chem. Lett. *3*, 61 (1967)
111. House, D. A., Garner, C. S.: Nature *208*, 776 (1965)
112. Wiede, O. F.: Chem. Ber. *32*, 378 (1899)
113. Fergusson, J. E., Wilkins, C. J., Young, J. F.: J. Chem. Soc. 2136 (1962)
114. Bhatnagar, S. S., Prakash, B., Hamid, A.: ibid. 1428 (1938)
115. Griffith, W.: ibid. 3948 (1962)
116. Werner, A.: Chem. Ber. *43*, 2286 (1910)
117. Hughes, R. G., Garner, C. S.: Inorg. Chem. *7*, 74 (1968)
118. Hughes, R. G., Ebsworth, E. A. V., Garner, C. S.: ibid. *7*, 882 (1968)
119. Stomberg, R.: Arkiv Kemi *22*, 49 (1964)
120. Stomberg, R.: ibid. *24*, 47 (1965)
121. Stomberg, R.: ibid. *23*, 401 (1965); Nature *205*, 71 (1965)
122. McLaren, E. H., Helmholz, L.: J. Phys. Chem. *63*, 1279 (1959)
123. Bartlett, B. L., Quane, D.: Inorg. Chem. *12*, 1925 (1973)
124. Peters, J. W., Pitts, Jr., J. N., Rosenthal, I., Fuhr, H.: J. Amer. Chem. Soc. *94*, 4348 (1972)
125. Stomberg, R., Brosset, C.: Acta Chem. Scand. *14*, 441 (1960)
126. Stomberg, R.: ibid. *17*, 1563 (1963)
127. Swalen, J. D., Ibers, J. A.: J. Chem. Phys. *37*, 17 (1962)
128. Dacre, P. D., Elder, M.: J. Chem. Soc. Dalton Trans. 1426 (1972)
129. Griffith, W. P.: J. Chem. Soc. 5345 (1963)
130. Klemm, W., Werth, H.: Z. Anorg. Allgem. Chem. *216*, 127 (1933)
131. Rodriguez Rios, B., Hoarau, J.: Compt. Rend. *241*, 1299 (1955)
131a. McGarvey, B. R.: J. Chem. Phys. *37*, 2001 (1962)
132. Von Wartenberg, H.: Z. Anorg. Allgem. Chem. *247*, 135 (1941)
133. Huss, E., Klemm, W.: ibid. *262*, 25 (1950)
134. Clark, H. C., Sadana, Y. N.: Canad. J. Chem. *42*, 50 (1964)
135. Maier, C. G.: U.S. Bur. Mines Bull. 436 (1942)
136. Sime, R. J., Gregory, N. W.: J. Amer. Chem. Soc. *82*, 93 (1960)
137. Oppermann, H.: Z. Anorg. Allgem. Chem. *359*, 51 (1968)
138. Shieh, C.-F., Gregory, N. W.: J. Phys. Chem. *77*, 2346 (1973)
139. Studer, P.: Diss. No. 740, Univ. Fribourg, 1975
140. Cuoni, B., Emmenegger, F. P., Rohrbasser, C., Schlapfer, C. W., Studer, P.: Spectrochim. Acta *34A*, 247 (1978)
141. Weber, J., Daul, C.: Mol. Phys. *39*, 1001 (1980)
142. Bode, H., Voss, E.: Z. Anorg. Allgem. Chem. *286*, 136 (1956)
143. Siebert, G., Hoppe, R.: Naturwiss. *58*, 95 (1971)
144. Sunder, W. A., Wayda, A. L., Distefano, D., Falconer, W., Griffiths, J. E.: J. Fluorine Chem. *14*, 299 (1979)
145. Allen, G. C., El-Sharkawy, G. A. M.: Inorg. Nucl. Chem. Lett. *6*, 493 (1970)
146. Hoffmann, B., Hoppe, R.: Z. Anorg. Allgem. Chem. *458*, 151 (1979)
147. Hoard, J. L., Williams, M. B.: J. Amer. Chem. Soc. *64*, 633 (1942)
148. Deadmore, D. L., Bradley, W. F.: Acta Cryst. *15*, 186 (1962)
149. Glemser, O., Roesky, H., Hellberg, K.-H.: Angew. Chem. Int. Ed. *2*, 266 (1963)
150. O'Donnel, T. A., Stewart, D. F.: Inorg. Chem. *5*, 1434 (1966)

151. Brown, S. D., Loehr, T. M., Gard, G. L.: J. Fluorine Chem. *7,* 19 (1976)
152. Edwards, A. J.: Proc. Chem. Soc. 205 (1963)
153. Edwards, A. J., Jones, G. R.: J. Chem. Soc. A, 1651 (1969)
154. Brown, S. D., Loehr, T. M., Gard, G. L.: J. Chem. Phys. *64,* 260 (1976)
155. Clark, H. C., Sadana, Y. N.: Canad. J. Chem. *42,* 702 (1964)
156. Green, P. J., Johnson, B. M., Loehr, T. M., Gard, G. L.: Inorg. Chem. *21,* 3562 (1982)
157. Krauss, H.-L., Munster, G.: Z. Naturforsch. *17b,* 344 (1962)
158. Johannesen, R. B., Krauss, H.-L.: Chem. Ber. *97,* 2094 (1964)
159. Seddon, E. A., Seddon, K. R., Thomas, V. H.: Trans. Met. Chem. *3,* 318 (1978)
160. Levason, W., Ogden, J. S., Rest, A. J.: J.C.S. Dalton Trans. 419 (1980)
161. Drew, M. G. B., Tomkins, I. B.: J. Chem. Soc., A, 22 (1970)
162. Doran, M., Hillier, I. H., Seddon, E. A, Seddon, K. R., Thomas, V. H., Guest, M. F.: Chem. Phys. Lett. *63,* 612 (1979)
163. Doran, M., Hawksworth, R. W., Hillier, I. H.: J.C.S. Faraday Trans., II, *76,* 164 (1980)
164. Krauss, H.-L., Leder, M., Münster, G.: Chem. Ber. *96,* 3008 (1963)
165. Banerjee, A. K., Banerjee, N.: Inorg. Chem. *15,* 488 (1976)
166. Seddon, K. R., Thomas, V. H.: J.C.S. Dalton Trans. 2195 (1977)
167. Sharpe, A. G., Woolf, A. A.: J. Chem. Soc. 798 (1951)
168. Nyholm, R. S., Sharpe, A. G.: ibid. 3579 (1952)
169. Ziebarth, O. V., Selbin, J.: J. Inorg. Nucl. Chem. *32,* 849 (1970)
170. Seddon, K. R., Thomas, V. H.: Inorg. Chem. *17,* 749 (1978)
171. Sarkar, S., Singh, J. P.: J.C.S. Chem. Comm. 509 (1974)
172. Gahan, B., Garner, D. C., Hill, L. H., Mabbs, F. E., Hargrave, K. D., McPhail, A. T.: J.C.S. Dalton Trans. 1726 (1977)
173. Brown, D.: J. Chem. Soc. 4944 (1964)
174. Wendling, E., Rohmer, R.: Bull. Soc. Chim. France 8 (1967)
175. Gray, H. B., Hare, C. R.: Inorg. Chem. *1,* 363 (1962)
176. Hare, C. R., Bernal, I., Gray, H. B.: ibid. *1,* 831 (1962)
177. Kon, H.: J. Inorg. Nucl. Chem. *25,* 933 (1963)
178. Kon, H., Sharpless, N.: J. Chem. Phys. *42,* 906 (1965)
179. Manoharan, P. T., Rogers, M. T.: ibid. *49,* 5510 (1968)
180. Garner, C. D., Miller, I. H., Mabbs, F. E.: Chem. Phys. Lett. *32,* 224 (1975)
181. Garner, C. D., Kendrick, J., Lambert, P., Mabbs, F. E., Hillier, I. H.: Inorg. Chem. *15,* 1287 (1976)
182. Weber, J., Garner, C. D.: ibid. *19,* 2206 (1980)
183. Sunil, K. K., Harrison, J. F., Rogers, M. T.: J. Chem. Phys. *76,* 3078 (1982): *76,* 3087 (1982)
184. Dalton, L. A., Bereman, R. D., Brubaker, Jr., C. H.: Inorg. Chem. *8,* 2477 (1969)
185. Garner, C. D., Hillier, I. H., Mabbs, F. E., Taylor, C., Guest, M. F.: J.C.S. Dalton Trans. 2258 (1976)
186. Garif'yanov, N. S.: Dokl. Akad. Nauk SSSR. *155,* 385 (1964)
187. Ryabchikov, D. I., Marov, I. N., Dubrov, Y. N., Belyaeva, V. K., Ermakov, A. N.: Dokl. Akad. Nauk SSSR *169,* 1107 (1966)
188. Verbeek, J. L., Cosnaz, P. F.: Recl. Trav. Chim. *86,* 209 (1967)
189. Hagihara, N., Yamazaki, H.: J. Amer. Chem. Soc. *81,* 3160 (1959)
190. Basi, J. S., Bradley, D. C.: Proc. Chem. Soc. 305 (1963)
191. Krauss, H.-L., Münster, G.: Z. Anorg. Allgem. Chem. *352,* 24 (1967)
192. Alyea, E. C., Basi, J. S., Bradley, D. C., Chisholm, M. H.: Chem. Comm. 495 (1968)
193. Alyea, E. C., Basi, J. S., Bradley, D. C., Chisholm, M. H.: J. Chem. Soc. A, 772 (1971)
194. Bradley, D. C., Hillyer, M. J.: Trans. Faraday Soc. *62,* 2382 (1966)
195. Dyrkacz, G., Roček, J.: J. Amer. Chem. Soc. *95,* 4756 (1973)
196. Bochmann, M., Wilkinson, G., Young, G. B., Hursthouse, M. B., Abdul Malik, K. M.: J.C.S. Dalton Trans. 1863 (1980)
197. Willis, C. J.: J.C.S. Chem. Comm. 944 (1972)
198. Bramman, P F., Lund, T., Raynor, J. B., Willis, C. J.: J.C.S. Dalton Trans. 45 (1975)
199. Basi, J. S., Bradley, D. C., Chisholm, M. H.: J. Chem. Soc. A, 1433 (1971)
200. Chisholm, M. H., Cotton, F. A., Extine, M. W., Rideout, D. C.: Inorg. Chem. *17,* 3536 (1978)

201. Mowat, W., Shortland, A. J., Hill, N. J., Wilkinson, G.: J.C.S. Dalton Trans. 770 (1973)
202. Yagupsky, G., Mowat, W., Shortland, A., Wilkinson, G.: J.C.S. Chem. Comm. 1369 (1970)
203. Mowat, W., Shortland, A., Yagupsky, G., Hill, N. J., Yagupsky, M., Wilkinson, G.: J.C.S. Dalton Trans. 533 (1972)
204. Mowat, W., Wilkinson, G.: J. Organomet. Chem. *38,* C35 (1972)
205. Kruse, W.: ibid. *42,* C39 (1972)
206. Müller, J., Holzinger, W.: Angew. Chem. Int. Ed. *14,* 760 (1975)
207. Bower, B. K., Tennet, H. G.: J. Amer. Chem. Soc. *94,* 2512 (1972)
208. Bochmann, M., Wilkinson, G., Young, G. B.: J.C.S. Dalton Trans. 1879 (1980)
209. Cardin, C. J., Cardin, D. J., Roy, A.: J.C.S. Chem. Comm. 899 (1978)
210. Tsuda, T., Kochi, J. K.: Bull. Chem. Soc. Jpn. *45,* 648 (1972)
211. Gramlich, V., Pfefferkorn, K.: J. Organomet. Chem. *61,* 247 (1973)
212. Cardin, C. J., Cardin, D. J., Kelly, J. M., Norton, R. J., Roy, A., Hathaway, B. J., King, T. J.: J.C.S. Dalton Trans. 671 (1983)
213. Ward, G. A., Kruse, W., Bower, B. K., Chien, J. C. W.: J. Organomet. Chem. *42,* C43 (1972)
214. Ward, G. A., Bower, B. K., Findlay, M., Chien, J. C. W.: Inorg. Chem. *13,* 614 (1974)
215. Krumpolc, M., Roček, J.: J. Amer. Chem. Soc. *101,* 3206 (1979)
216. Krumpolc, M., DeBoer, B. G., Roček, J.: ibid. *100,* 145 (1978)
217. Roček, J., Peng, T. Y.: ibid. *99,* 7622 (1977)
218. Srinivasan, K. G., Roček, J.: ibid. *100,* 2789 (1978)
219. Ramesh, S., Mahapatro, S. N., Liu, J. H., Roček, J.: ibid. *103,* 5172 (1981)
220. Wiberg, K. B., Mukherjee, S. K.: ibid. *96,* 6647 (1974)
221. Srinivasan, V. S., Gould, E. S.: Inorg. Chem. *20,* 3176 (1981)
222. Rajasekar, N., Gould, E. S.: ibid. *22,* 3798 (1983)
223. Kadish, K., Davis, D. G., Fuhrhop, J.-H.: Angew. Chem. Inter. Ed. *11,* 1014 (1972)
224. Nill, K. H., Wasgestian, F., Pfeil, A.: Inorg. Chem. *18,* 564 (1979)
225. Budge, J. R., Gatehouse, B. M. K., Nesbit, M. C., West, B. O.: J.C.S. Chem. Comm. 370 (1981)
226. Buchler, J. W., Lay, K. L., Castle, L., Ullrich, V.: Inorg. Chem. *21,* 842 (1982)
227. Groves, K. T., Kruper, Jr., W. J., Haushalter, R. C., Butler, W. M.: ibid. *21,* 1363 (1982)
228. Groves, J. T., Haushalter, R. C.: J.C.S. Chem. Comm. 1165 (1981)
229. Groves, J. T., Kruper, Jr., W. J.: J. Amer. Chem. Soc. *101,* 7613 (1979)
230. Matsuda, Y., Yamada, S., Murakami, Y.: Inorg. Chim. Acta *44,* L309 (1980)
231. Buchler, J. W., Dreher, C., Lay, K.-L., Raap, A., Gersonde, K.: Inorg. Chem. *22,* 879 (1983)
232. Groves, J. T., Takahashi, T., Butler, W. M.: ibid. *22,* 884 (1983)
233. Buchler, J. W., Dreher, C., Lay, K.-L.: Chem. Ber. *117,* 2261 (1984)
234. Macor, K. A., Spiro, T. G.: J. Amer. Chem. Soc. *105,* 5601 (1983)
235. Murakami, Y., Matsuda, Y., Yamada, S.: J.C.S. Dalton Trans. 855 (1981)
236a. Davison, A., Edelstein, N., Holm, R. H., Maki, A. H.: ibid. *86,* 2799 (1964)
236b. Waters, J. H., Williams, R., Gray, H. B., Schrauzer, G. N., Finck, H. W.: ibid. *86,* 4198 (1964)
236c. Brunner, H., Wachter, J., Guggolz, E., Ziegler, M. L.: ibid. *104,* 1765 (1982)
237. Girolami, G. S., Salt, J. E., Wilkinson, G., Thornton-Pett, M., Hursthouse, M. B.: ibid. *105,* 5954 (1983)
238. Arzhankov, S. I., Poznyak, A. L.: Russ. J. Inorg. Chem. *26,* 850 (1981)
239. Arzhankov, S. I., Poznyak, A. L.: Z. Anorg. Allgem. Chem. *481,* 201 (1981)
240. Siddall, T. L., Miyaura, N., Huffman, J. C., Kochi, J. K.: J.C.S. Chem. Comm. 1185 (1983)
241. Sulfab, Y., Al-Shatti, N. I., Hussein, M. A.: Inorg. Chim. Acta *86,* L59 (1984)
242. Jennette, K. W.: J. Amer. Chem. Soc. *104,* 874 (1982)
243. Goodgame, D. M. L., Hayman, P. B., Hathaway, D. E.: Polyhedron *1,* 497 (1982)
244. Kazansky, P., Pacherskaya, V.: Kinet. Catal. USSSR *2,* 417 (1961): *3,* 315 (1962): *4,* 210 (1963): *5,* 379 (1964)
245. van Reijen, L., Cossee, P.: Dissc. Faraday Soc. *41,* 277 (1966)
246. Lunsford, J. H.: Catal. Rev. *12,* 137 (1975)
247. Hoskins, R. H., Soffer, B. H.: Phys. Rev. *133,* 499 (1964)
248. Pontnau, J., Adde, R.: ibid. *B14,* 3778 (1976)
249. Pontnau, J., Adde, R.: J. Phys. Chem. Solid *36,* 1023 (1975)

250. Ludwig, G. W., Woodbury, H. H.: Solid State Phys. *13,* 265, 271 (1962)
251. Tsukioka, M., Yamamoto, A., Kojima, H.: J. Phys. Soc. Jpn. *33,* 681 (1972)
252. Kurkin, I. N., Chirkin, Yu. K., Shlekin, V. I.: Sov. Phys. Solid State *17,* 2028 (1975)
253. Kedzie, R. W., Shane, J. R., Kestigian, M.: Phys. Lett. *11,* 286 (1964)
254. Lingam, K. V., Nair, P. G., Venkatraman, B.: Proc. Ind. Acad. Sc. *70A,* 29 (1969)
255. Gourier, D., Parant, J. P., Vivien, D.: Phys. Status Solidi *B113,* 509 (1982)
256. Ravi Sekhar, Y., Bill, H.: Chem. Phys. Lett. *92,* 349 (1982)
257. Manjunath, C. V., Srinivasan, R.: Phys. Status Solidi *B87,* 395 (1978)
258. Lister, D. H., Symons, M. C. R.: J. Chem. Soc. A, 782 (1970)
259. Debuyst, R., Apers, D. J., Capron, P. C.: J. Inorg. Nucl. Chem. *34,* 1541 (1972)
260. Grunin, V. S., Patrina, I. B., Zonn, Z. N.: Sov. Phys. Solid State *22,* 2045 (1980)
261. Lagendijk, A., Morel, R. J., Glasbeek, M., Van Voorst, J. D. W.: Chem. Phys. Lett. *12,* 518 (1972)
262. de Jong, H. J., Glasbeek, M.: Solid State Comm. *19,* 1197 (1976): *28,* 683 (1978)
263. Glasbeek, M., de Jong, H. J., Koopmans, W. E.: Chem. Phys. Lett. *66,* 203 (1979)
264. Grunin, V. S., Patrina, I. B., Zonn, Z. N.: Phys. Status Solidi *B98,* 765 (1980)
265. Alybakov, A. A., Toichiev, N.: ibid. *B101,* 597 (1980)
266. Howard, D. G., Lindquist, R. H.: J. Chem. Phys. *38,* 573 (1963)
267. Deren, J., Haber, J., Kosek, S.: Bull. Akad. Pol. Sc. Ser. Chem. *13,* 21 (1965)
268. Mitewa, M., Bontchev, P. R.: Coord. Chem. Rev. *61,* 241 (1985)
269. Chamberland, B. L., Herrero-Fernandez, M. P., Hewston, T. A.: J. Solid State Chem. *59,* 111 (1985)
270. Hope, E. G., Jones, P. J., Levason, W., Ogden, J. S., Tajik, M., Truff, J. W.: J.C.S. Dalton Trans. 1443 (1985)
271. Bongon, R., Wilson, W. W., Christe, K. O.: Inorg. Chem. *24,* 2286 (1985)
272. Jacob, E. J., Hedberg, L., Hedberg, K., Davis, H., Gard, G. L.: J. Phys. Chem. *88,* 1935 (1984)
273. Hope, E. G., Jones, P. J., Levason, W., Ogden, J. S., Tajik, M., Turff, J. W.: J.C.S. Dalton Trans. 2445 (1984)
274. Roch, M., Weber, J., Williams, A. F.: Inorg. Chem. *23,* 4571 (1984)
275. Compochiaro, C., Hoffmann Jr., J. A., Bocian, D. F.: ibid. *24,* 449 (1985)
276. Liston, D. J., West, B. O.: ibid. *24,* 1568 (1985)
277. Liston, D. J., Kennedy, B. J., Murray, K. S., West, B. O.: ibid. *24,* 1561 (1985)
278. Abrahamson, H. B., Brandenburg, K. L., Lucero, B., Martin, M. E., Dennis, E.: Organometallics *3,* 1379 (1984)

Author-Index Volumes 1–63

MIX
Papier aus verantwortungsvollen Quellen
Paper from responsible sources
FSC® C105338

If you have any concerns about our products,
you can contact us on
ProductSafety@springernature.com

In case Publisher is established outside the EU,
the EU authorized representative is:
Springer Nature Customer Service Center GmbH
Europaplatz 3, 69115 Heidelberg, Germany

Printed by Libri Plureos GmbH
in Hamburg, Germany